Beginner's
Guide to
Electronics

D1633867

Beginner's Guides are available on the following subjects:

Audio
Building Construction
Cameras
Central Heating
Colour Television
Computers
Digital Electronics
Domestic Plumbing
Electric Wiring
Electronics
Fabric Printing and Dyeing
Gemmology
Home Energy Saving
Integrated Circuits
Microprocessors
Photography
Processing and Printing
Radio
Spinning
Super 8 Film Making
Tape Recording
Technical Illustration
Television
Transistors
Video
Weaving
Woodturning
Woodworking

Beginner's Guide to **Electronics**

4th Edition

Owen Bishop

B.Sc. (Bristol)., B.Sc. (Oxon).

Newnes Technical Books

Published by Newnes Technical Books
an imprint of Newnes Books
a division of The Hamlyn Publishing Group Limited
84-88 The Centre, Feltham, Middlesex, TW13 4BH,
and distributed for them by
Hamlyn Distribution Services
Rushden, Northants, England

First published 1964 by George Newnes Ltd.
Second edition 1967
Third edition 1974 by Newnes Technical Books
Reprinted 1975, 1977, 1978, 1980
Fourth edition 1982
Reprinted 1983, 1984, 1985

British Library Cataloguing in Publication Data

Bishop, Owen
Beginner's guide to electronics. - 4th edition
1. Electronic apparatus and appliances
I. Title   II. Squires, Terence Leighton
Beginner's guide to electronics
621.381   TK7870

ISBN 0-408-00413-4

Typeset by Butterworths Litho Preparation Department
Printed and Bound in Great Britain by Mackays of Chatham Ltd

Preface

Electronics now plays an increasingly important part in all walks of life. Electronic techniques are the basis of most modern scientific research; they make possible much of our everyday entertainment, enable industrial processes from the simplest to the most elaborate to be controlled automatically, play a vital part in air and sea navigation, and are essential to modern defence and warfare. Such a subject is bound to excite the interest of intelligent young people of a practical or scientific turn of mind.

This book has been written to help those who are thinking of starting a career in electronics. It is written also for those who do not wish to be mere button-pushers, but who wish to be able to understand something of the technology that has made such a vast difference to our lives. It assumes no prior technical knowledge on the part of the reader. The subject is dealt with without recourse to mathematics, emphasis being placed on illustrative diagrams so that an understanding of this complex subject can be gained rapidly.

When the third edition was being prepared, less than a decade ago, electronics was still in transition from the age of the vacuum tube to the age of the semiconductor. Equipment based on thermionic valves was operating alongside equipment containing the newly developed integrated circuits. Now the transition is virtually complete. The whole emphasis of this book has been changed to reflect that transition. With the rapid advances in solid-state technology, new families of electronic devices have been developed and additional space has been given to them. There have been developments in the use of microwaves, and the chapter on this subject has been expanded. To reflect the more widespread use of tape-recorders both for audio and video recording, a new chapter on recording has been added.

It has been held by some people that the development of integrated circuits has made electronics less interesting and less attractive as a hobby. A glance at the hobby magazines shows this view to be unfounded. Now the amateur uses integrated circuits as the building-blocks of electronic equipment that earlier generations would have thought impossible to construct. This book deals at some length with integrated circuits, in the hope that it will encourage those new to the subject to take it up for the fascinating hobby that it is.

In general, it is hoped that the reader will find much between these covers that is helpful and informative.

Wiseton   O.B.

Contents

1 Electric Currents

The story of electronics began in 1883 when the American physicist and engineer Thomas A. Edison discovered that under certain conditions electricity will flow through a vacuum. He had been experimenting with a small metal plate placed inside an evacuated electric light bulb. When he made this plate electrically positive with respect to the filament of the bulb, he found that an electric current flowed through the vacuum between the filament and the plate, that is through the lamp. But before we consider the significance of this simple experiment, let us consider what an electron is and the laws it obeys.

When we talk about matter we usually mean nearly everything that we can detect by our senses. A simple definition of matter would be anything that occupies space and has weight. Matter can exist in any of the three conditions: solid (or frozen), liquid or gaseous. Examples of these conditions (at normal temperature and pressure) are iron, water and air. At normal room temperature iron is frozen, that is solid, but if the temperature is raised to 1000°C then it becomes molten or liquid. At room temperature water is molten or liquid whereas at 0°C it becomes frozen or solid. At nearly 200°C below zero air becomes liquid and at a few degrees lower still it becomes solid.

Generally speaking gases can be made into liquids or solids by changing their temperature and/or pressure, and liquids can be made gaseous or solid. In the case of solid matter, however, there are many complex types which can exist only as solids.

Molecules

If matter is broken down by chemical techniques it will be found to consist of molecules. For example, wood consists of molecules of resin and cellulose. Air consists of molecules of oxygen, nitrogen, argon, carbon dioxide and a few rarer gases.

Molecules in turn are composed of one or more of the atoms of the hundred or so elements known to science. Most of these elements may be found in any of the three states previously mentioned. The element oxygen's form up in two's to produce a simple oxygen molecule which is gas. Iron and copper molecules are simply atoms of these elements held in a crystal structure forming a solid. Polytetrafluoroethylene, on the other hand, a solid much used as an insulator in electronics, is a complicated molecule made up of atoms of the elements carbon and fluorine.

Simple molecules are common salt, which consists of only two atoms, one each of sodium and chlorine, and copper, in which the molecule and the atom are the same unit. A molecule of water (chemical abbreviation H_2O) consists of two atoms of hydrogen (H_2) and an atom of oxygen (O).

Electrons

An atom is the smallest unit which still retains the identity of an element. If the atom is broken down still further one obtains a collection of electrons, protons and other particles, many of which have only recently been discovered by scientists.

From the electronics point of view only two particles need concern the reader: electrons, and the remainder of the atom which is called the nucleus.

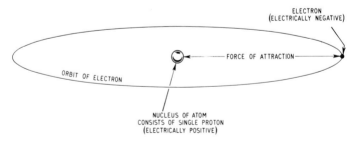

Fig. 1.1. The hydrogen atom

Of the hundred or so known elements, the hydrogen atom has the simplest structure, consisting of a single electron moving in orbit around a nucleus which is a single proton. This is illustrated in Fig. 1.1.

To most people a millimetre is a small unit of measurement, although in engineering, precisions of ±0.01 mm are sometimes required. However, on an atomic scale even this is an enormous measurement. The hydrogen atom is only about one ten-millionth of one millimetre in diameter, and an electron is a mere one ten-thousandth of the size of an atom. If the atom were scaled up so that the nucleus was 1 mm in diameter, the electron would be in orbit about 120 metres away from the nucleus. Thus one can see that the atom is composed mainly of space.

Electricity

The electron has a negative charge: it is always attracted to any body which has a positive charge. In the nucleus of an atom are particles called protons which have a positive charge, and in normal circumstances this positive charge is equal to the negative charge of the atom's electrons. If an atom has fifty (negative) electrons then there will be fifty (positive) protons in the nucleus so that the electrical charges balance, for in its normal state the atom is always electrically neutral. If an atom loses an electron it has an excess of positive electricity and is called a *positive ion*. If it gains an electron it becomes a *negative ion*. The movement of electrons and ions under various influences forms the basis of all work in electronics.

Electrons orbit the nucleus at a number of different distances. Not more than two electrons can orbit at the smallest distance shown in Fig. 1.1. The atom of helium, with 2 protons in its nucleus, has 2 electrons orbiting at the same distance as in the hydrogen atom. Lithium, with 3 protons has 3 electrons and, since there is room for only 2 at the smallest orbiting distance, the third electron orbits at a slightly greater distance. Up to 2 electrons can orbit at this distance, so with beryllium (4 protons) this next orbiting region, or *shell,* is full. Boron has 5 electrons, 2 in the innermost shell, 2 in the next shell and one in the outer shell. Shells outside these may contain up to 2, 6, 10 or 14 electrons, depending on the shell. In most elements the shells are filled in order from the innermost outwards. For example, copper (29 protons) has six full shells containing 2, 2, 6, 2, 6 and 10 electrons, plus 1 electron in the outermost shell. This electron is available for conducting electricity, as will be explained later.

Current

An electric current is the organized movement of electrons in a material, the number of electrons flowing past a certain point in a given time being the rate of current flow. Electric current (symbol I) is measured in amperes, milliamperes (1/1000 of an ampere) or microamperes (one millionth of an ampere), the last two units being the ones most commonly used in electronics. A microampere is equal to about six billion electrons flowing past a point in a second.

The actual number of electrons that flow—irrespective of time—is measured in coulombs. A 'box' containing a coulomb of electricity would store over six times a million, million, million electrons.

Conventional current

Like many other sciences, the study of electricity began to yield practical results before a basic theory to explain electrical phenomena could be formulated. Early investigators thought that electric currents flowed from the positive terminal of a source of electrical energy, for example a battery, to the negative terminal (in the external circuit and not within the battery itself). This has come to be called the direction of *conventional current flow*. Today we know that the movement of the electrons, i.e. the current flow, is from the negative to the positive terminal.

The electronic engineer must remember this when talking about the direction of the current as he may be asked to specify the system he is talking about.

Conductors and insulators

Electric currents will pass more readily through some materials than others. Materials which readily pass electric currents are termed conductors while those that resist the passage of an electric current are known as insulators.

The atoms of some elements possess electrons which form an outer shell only weakly bound to the nucleus. Copper is an example, and is a commonly used conductor. As mentioned above, its outer shell contains a single electron. Its real significance as a conductor, however, is its behaviour when a large number of atoms are grouped together to form a small block of the material. The atoms form a matrix or regular structure and remain linked in this fashion until the material is destroyed by physical or chemical means. While the atoms remain fixed in a matrix, as shown in Fig. 1.2, they become ionized; that is, some

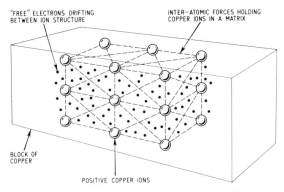

Fig. 1.2. Arrangement of atoms in a copper conductor

of their outer shell electrons detach themselves from the parent atom and drift about the lattice rather like pigeons among the ornamentation on city buildings. Under the right influence, these free electrons will form an electric current. Most common metals act as conductors.

There are, on the other hand, elements which have atoms whose electrons orbiting on outer shells are more tightly bound to the nucleus (because the shell is full or nearly full) and so are less ready to move about in the molecular structure. Materials that contain a high proportion of much elements are insulators, examples being pure water, dry paper, rubber, etc.

Table 1.1. CONDUCTIVITY OF VARIOUS MATERIALS

Materials that conduct electricity	Materials that resist the flow of current but still allow some to occur	Materials that are good insulators and will not allow any current to flow through them
Silver	Iron	Rubber
Gold	Oxides of metals	Bakelite
Copper	Carbon	Polytetrafluoroethylene
Brass	Nickel-chrome	Polystyrene
Platinum	Germanium*	Ebony
Tin	Silicon*	Wood
Lead	Lead sulphide	Dry paper
Niobium		Ceramic
Palladium		Pure water
Molybdenum		Mica
Saline solution		Asbestos
(salt water)		Air

* These elements occupy a special category known as semiconductors.
  Note that many conductors occur naturally as elements whereas many insulators are complex man-made structures.

An electric current flows easily in a conductor, it flows with some difficulty in several 'in-between' materials, and negligibly or not at all in an insulator.

Conductors vary to some extent in their 'willingness' to conduct, for example, silver allows a greater current to flow under the same conditions than copper or iron.

Why an electric current flows in a conductor – an e.m.f.

An electron current is formed by the migration of electrons under some external influence such as a cell as shown in Fig. 1.3 But what happens inside the cell? Inside the cell the circuit is completed by a paste or liquid electrolyte and two electrodes as shown in Fig. 1.4. The electrodes are made of conducting

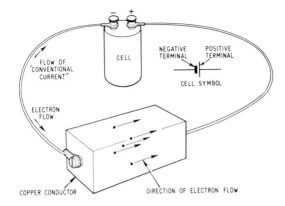

Fig. 1.3. Current flow in a conductor

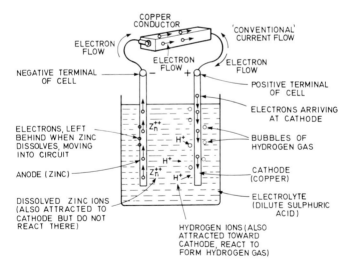

Fig. 1.4. Producing an e.m.f. from a simple cell

material, such as zinc, copper or carbon. The electrolyte consists of a salt solution, though in 'dry' cells it may also contain materials to make it into a sticky paste. The dissolved salts split into ions, which are charged. For example, when the ammonium chloride in a dry cell ionizes, its molecules split to form an ammonium ion and a chlorine ion. The ammonium ion loses one electron in the process, becoming positively charged, and the chlorine gains this electron, becoming negatively charged. Fig. 1.4 shows the ions formed in dilute sulphuric acid. Since ions are charged and are mobile some of them can carry electric charge across from one electrode to the other.

To see what happens next, consider the action of the simple cell of Fig. 1.4. One of the electrodes is zinc, and some of its atoms dissolve in the electrolyte, forming zinc ions which are positively charged (cations). This leaves the zinc electrodes with a negative charge. It becomes the anode. This process continues until the negative charge becomes great enough on the anode to begin to attract the positively charged zinc ions. Equilibrium is reached, with the anode having a slightly higher potential than the electrolyte and no more zinc dissolves. The electrical potential is the result of conversion of chemical energy by the chemical action of forming a solution.

The other electrode is copper and has different properties: it does not dissolve in the electrolyte. There is a tendency for copper to attract positive ions out of solution, giving the electrode a positive charge. In the cell this does not happen until the electrodes are connected by a conductor or some other external circuit, when the positively charged hydrogen ions flow to the copper cathode. There they are discharged by electrons that have come from the anode by way of the external circuit. The discharged ions form bubbles of hydrogen gas. Now that the surplus electrons have gone from the anode, its negative charge is reduced, and it no longer attracts the zinc ions so strongly; more zinc can dissolve, releasing more electrons at the anode. This keeps up the flow of current. The process continues until the zinc has more-or-less completely dissolved away, and the electrolyte has become a solution of zinc sulphate. Other types of cell have different electrolytes and different pairs of conducting elements (carbon/iron, nickel/cadmium) as electrodes, but the principle of the conversion of chemical energy to electrical energy is essentially the same in all. The conversion generates a force – the *electromotive force,* or e.m.f. for short — which drives electrons around the external circuit and causes a flow of ions across the cell.

This electromotive force is measured in a unit called a volt. In electronics the millivolt (one thousandth of a volt, abbreviation mV) and the microvolt (one millionth, abbreviation μV) are the most common units to be employed. The simple cell just described is capable of providing an output voltage of about 1.1 volts.

Relationship between e.m.f. and current — resistance

It is the e.m.f. that makes the electrons flow in a conductor. How many electrons flow in a particular conductor, that is, how large the current is, depends first upon how great is the e.m.f. and secondly on the resistance to current flow offered by the conductor. For a conductor of a given resistance, doubling the e.m.f. doubles the current flowing.

If we make a conductor of, for example, a mixture of copper dust and graphite powder the current that will flow will be reduced below the amount that would flow when pure copper is used even though the e.m.f. is kept at the same value. This is because the second material, graphite, is a poorer conductor–it resists the flow of current by not allowing enough electrons to take part in the action.

In practice, at normal temperatures all conductors have some resistance. However, there are some materials known as super conductors that lose their resistance completely when cooled below a certain critical temperature, characteristic of the material. The critical temperature is always near absolute zero, which is −273°C or 0 K (K stands for Kelvin and is the unit of the absolute temperature scale. Its degrees have the same magnitude as those of the Celsius scale). For example, lead becomes superconducting below 7.2 K and cadmium only becomes superconducting below 0.54 K. In experiments to determine whether superconductors do possess any resistance at all, currents have been set up in closed loops and have flowed undiminished for more than a year.

At present, considerable research is going on to produce materials that exhibit superconductivity at normal temperatures.

The symbol for resistance is shown in Fig. 1.5. A good conductor can be regarded as a theoretically perfectly conducting wire with a hypothetical resistance in series, although this resistance will have a very small value.

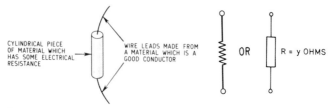

CYLINDRICAL PIECE OF MATERIAL WHICH HAS SOME ELECTRICAL RESISTANCE

WIRE LEADS MADE FROM A MATERIAL WHICH IS A GOOD CONDUCTOR

OR

R = y OHMS

Fig. 1.5. Appearance of a typical small resistor and the symbol used in circuit diagrams to denote a resistor

Resistance (R) is measured in ohms (symbol Ω), but in electronics we may be often be dealing with megohms (millions of ohms, MΩ) and kilohms (thousands of ohms, kΩ).

To summarise, if the e.m.f. is increased, the current flow is increased; if the resistance is increased, the current flow decreases.

Combinations of resistors

If resistors are connected together in parallel the effective resistance is reduced, while if they are connected together in series the resistance is increased. This can be illustrated by an analogy. It is easier to empty a football stadium if you have many exits and it is easier for the e.m.f. to produce a larger current in a circuit if it has a number of parallel paths to drive the current through as seen in Fig. 1.6. Conversely, if the stadium exits were all placed in series it might take a whole weekend to empty it and, in the same way, if resistors are placed in series the total

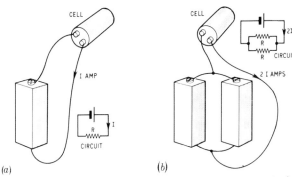

Fig. 1.6. Effect of parallel paths on current flow. (a) A single path through one conductor which has a small amount of resistance, R ohms. (b) The effect of two conducting paths on the current flowing. Both conductors have the same resistance value (R ohms) as the conductor in case (a), thereby allowing twice the amount of current to flow. (Two resistors in parallel means a smaller total resistance)

resistance presented to or 'seen' by the e.m.f. is greater and the resultant current flow is less. By the same reasoning if a conductor, say a wire x feet long, has a resistance of y ohms and its length is increased to $2x$ feet then the resistance would be increased to $2y$ ohms. If E volts were forcing a current I amperes through the first wire, the same e.m.f. in the second case would produce only $I/2$ amperes.

Sources of e.m.f.

There are of course other sources of e.m.f. besides cells. As everyone knows, coal is burnt to produce steam and mechanical power which will drive a large generator, the electrical output of which is widely distributed to various users. In places distant from such a mains supply a petrol-driven or wind-driven dynamo can be employed. (The Russians are using wind generators mounted on balloons six miles above the earth in what is known as the tropopause layer.) Hydroelectricity is another important source of electrical energy. Here rivers are dammed and the energy of the arrested water used to drive turbogenerators. The tidal motion of the sea is in use at the Rance tidal estuary power station in France, and many new devices for utilising tidal power are being developed.

Today, nuclear power is being used as a source of electrical energy. The coolant used to carry away exothermic heat from a nuclear reaction can be used instead of coal to produce steam.

Ten years ago batteries would have received scant attention in an electronics textbook since their use was limited to a few portable radio receivers. Today, however, with the immense growth of electronics and the wide use of transistor devices, the battery is once more important as a source of e.m.f.

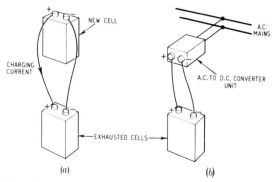

Fig. 1.7. Recharging an exhausted cell (secondary type). Method (a), using a new cell; method (b), normal method of recharging from the mains supply

Cells are of two kinds: primary cells and secondary cells. The former exhaust themselves and have to be thrown away after some time in use. Secondary cells can be recharged, after they have exhausted their chemical energy, by reversing the direction of the current through them as shown in Fig. 1.7.

Recharging could be done by means of another fully charged battery but in practice it is more likely to be done from a source derived from mechanical energy, such as the mains power supply.

Life of a cell

A cell supplies an e.m.f. at a value fixed by the chemical and physical characteristics of the cell. This e.m.f. will drive a current through an external circuit for a period of time which also is determined by these characteristics. The capacity of a cell is calculated in ampere-hours. For example, a 10 ampere-hour cell will supply, at a fixed e.m.f., a current of 10 amperes for 1 hour or 10 milliamperes for about 40 days (1000 hours).

The miniature cells used in pocket calculators and electronic watches have an e.m.f. of 1.4V and capacity of 35 to 100 milliampere-hours (1mA for 35 to 100 hours or 35 to 100mA for 1 hour), depending on type.

Whatever the source of the e.m.f., it is apparent that in order to get electrical energy some other form of energy, mechanical or chemical, has to be drawn upon.

Potential difference

The amount of the e.m.f. present between the positive and negative terminals of a source of e.m.f. is termed the potential difference (p.d.). Like e.m.f., p.d. is expressed in volts. In Fig. 1.8. an e.m.f. derived from a cell is shown driving a current

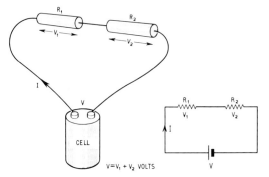

Fig. 1.8. Potential difference

through two resistors connected in series. The e.m.f. can be said to be 'shared' by the two resistors, some of it 'appearing across' each one. The amount that appears across each is known as the potential difference across the resistor.

If one of the resistors has a large resistance value and the other a small value most of the e.m.f. will appear across the larger value resistor, a proportionately smaller amount appearing across the smaller value resistor.

Ohm's law

The mathematical relationship between potential difference, current flow and the resistance in an electric circuit was first investigated by the scientist Georg Ohm and is called Ohm's law after him (as also is the unit for resistance, the ohm). This relationship is

$$R = \frac{V}{I}$$

where R is the resistance in ohms, V the potential difference (volts) and I the current (in amperes). The formula may of course also be written $I = V/R$, or $V = RI$. Thus if we know two of these quantities we can always calculate the third one.

Power

When a poor conductor, for example nickel-chrome wire, is placed across the terminals of a cell a certain small current will flow. This may be say, one hundredth of the current that would have flowed if the conductor had been a very good one, such as a block of copper. To produce the same current in both of these cases, a larger e.m.f. would be needed for the nickel-chrome wire. This could be obtained by using a number of cells connected in series. We call such an arrangement of cells a *battery* of cells, or a 'battery,' for short. To produce a high current flow through a greater resistance a high e.m.f. is required and this means that more batteries are needed and therefore more chemical energy is converted into electrical energy. The rate at which this energy is used up is called the *power consumption* of the resisting element and is expressed in watts (abbreviation W) or milliwatts (one thousandth of a watt abbreviation mW). It is given by the product volts×amperes.

Where does the energy disappear to when a large current is forced through an electrical resistance? In poor conductors electron collisions occur in the molecular lattice structure and energy is expended in this process. If the nickel-chrome wire is felt when a current is passed through it, it will be noticed that the wire is warm. The electrical energy is being converted to heat energy during the passage of current through the resisting material. This phenomenon is of course what happens in the ordinary electric light bulb and electric fire, where their resistance and the current forced through them make the filaments in both cases radiate light and heat energy.

In an electric fire thousands of watts are dissipated in this way. By way of comparison, the input circuit of a special measuring instrument used in electronics may dissipate only a fraction of a picowatt (a millionth of a millionth of a watt).

Output voltage and internal resistance

A very important idea which is constantly in the mind of the practising electronics engineer is that of internal resistance. In Fig. 1.4. a simple cell was shown forcing a current through a conducting block of copper. The current flow would be fairly high under these circumstances but, although the copper block has negligible resistance, the cell has an 'internal resistance' which limits the current flow. The output voltage provided by the cell is also affected by this internal resistance, i.e. some of the e.m.f. is lost as a voltage or potential difference across the cell's internal resistance.

All sources of e.m.f. have an internal resistance–sometimes called the source resistance. In the case of a cell it can be seen from Fig. 1.4 that the current flowing in the external circuit also flows inside the cell–in fact it is the same current. The electrodes and the electrolyte will have some resistance, even though it may be small–less than an ohm. Thus when the cell delivers a current, some of the e.m.f. is lost as potential difference across this internal resistance. The higher the current flow, the greater the e.m.f. lost and the smaller the voltage appearing across the terminals. This is demonstrated in Fig. 1.9.

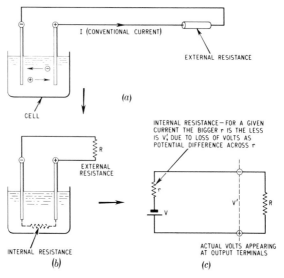

Fig. 1.9. The concept of internal resistance

Whatever the source of e.m.f.–motor generator, oscillator-power source or rectifying power supply, each has this 'built-in' resistance. In some cases it is high: this can be a nuisance or, on the other hand, a protection. In other cases it is very low so that large currents can be drawn from the unit while the output voltage remains constant. (The higher the internal resistance the greater the voltage fluctuation with changes in the current flowing–see Ohm's Law, given earlier in this chapter).

The generators that supply a town with electricity must have a very low internal resistance since they have to supply very heavy currents. Stabilized power supplies, now so commonly used in electronics, are also designed to have very low internal resistance because it is important to have a constant output voltage whatever value of current is drawn.

The concept of internal resistance—or as it is called in later chapters input or output impedance—is one of the most important in electronics. It is a factor which affects the design of many instruments and determines methods of measurement. For example, it is no use trying to measure accurately the output voltage of a circuit with a high internal resistance or impedance using a voltmeter which needs a considerable current to operate the movement. Were such a meter used, the output voltage of the circuit would fall as soon as the meter was placed across the points to be measured and thus a wrong reading would be obtained. In a case like this, a special measuring instrument—a FET-voltmeter—can be used to obtain an accurate reading of the circuit output voltage.

Matching

Another important concept which is always occurring in electronics is that of correct matching (see Fig. 1.10). Consider the simple case of a generator supplying power to an electric light bulb. The generator and lamp are said to be *matched* when the optimum amount of power is being transferred from the generator to the lamp. This does not mean that all the power produced by the generator is delivered to the load (i.e. the lamp). Some, which is dissipated in the internal resistance of the generator, is wasted, only the remainder of the power being delivered to the load. It is, therefore, desirable to obtain conditions whereby the maximum amount of power is transferred to the load. It can be proved algebraically and readily shown with simple laboratory equipment that these conditions obtain when the resistance of the load is equal to the internal resistance of the generator. Under these conditions the generator is said to be matched to the load.

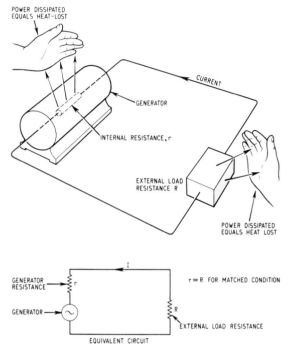

Fig. 1.10. The matched condition. When the generator and the external load are matched, hand detects the same heat from both

The expression 'matched' occurs frequently in electronics. For example, an oscillator oscillating at say 10 million hertz (10 MHz) is really a generator and it gives its best performance if its output impedance (i.e. internal resistance) is matched to the impedance of the load. Again, an audio amplifier operating at say 1000 Hz will deliver maximum power to a loudspeaker if the loudspeaker's resistance or impedance is equal to the output impedance of the amplifier. The power produced in the external load is at a maximum under these conditions.

Practical conductors

It is usual to use copper wire, either insulated with a protective sleeving or not according to the circumstances, to connect together components in an electronic circuit. The wire is usually supplied in reels or coils and can be obtained in various gauges. The usual way to specify the diameter of the wire is according to the standard wire gauge (s.w.g.). For most circuit work 20 s.w.g. wire has an appropriate thickness and pliability although where the wire is to be self-supporting a heavier gauge may be used, for example 16 s.w.g. Nearly all copper wire is tinned, that is covered with a coating of tin to assist in soldering the wire to the various components.

The gauge of wire can also be expressed by its diameter, in millimetres (mm), or its area of cross section, in square millimetres (mm^2).

In the case of wire with an insulating sleeving to protect it from accidental contact with other parts of the circuit the sleeving is usually made of p.v.c., an insulating plastic material. The sleeving is available in different colours so that if a section of a circuit is wired in a particular colour it can be easily identified. In much more modern electronic equipment 'printed circuits' are used.

Practical batteries

Different types of battery are available for different requirements. Secondary cells such as those used in motor cars have a capacity of about 80 ampere-hours, i.e. they can deliver up to 20 amperes for four hours without any appreciable change in their output voltage (or e.m.f.). This type of battery uses dilute sulphuric acid as its electrolyte, the electrodes taking the form of lead plates to provide a substantial surface area to the electrolyte –a factor which determines the electrical capacity of the battery. Each cell will provide an output voltage of 2 volts, this being determined by the electro-chemical nature of the cell. To obtain higher voltages a number of the cells can be connected in series, positive terminal to negative terminal and so on. If a number of the cells are connected in parallel, that is all the positive terminals together and all the negative ones together, the output voltage will remain the same as for one cell but the capacity will increase by a factor equal to the number of cells connected in this way. Secondary cells are very useful in laboratories where a large current at a constant voltage is needed.

Most of the smaller batteries consist of primary cells. With the use of transistors and semiconductor devices in modern equipment, ranging from medical instrumentation to guided missile telemetry, the use of small primary cells has greatly increased during the last few years. The electrolyte used is usually a paste of some kind, hence the term 'dry cell'. This type of cell can be physically very small and yet still provide a useful performance.

Standard cells are another important type. They are designed to give a constant output voltage which should not change over a temperature range of −30°C to +70°C by more than ±0.5 per cent, with a current drain on the cell of about 100 microamperes. These figures are obtained by designing a cell with a very low internal resistance. They are used by electronics engineers as reference standards when calibrating instruments, which may themselves be used later to make measurements on electronic and other types of laboratory apparatus.

Resistors

Resistors are made of a mixture of conducting and insulating material. In manufacture the proportions are adjusted to give resistances of various values.

Three types of composition resistor are commonly used: the lacquer type, which is employed mostly in variable resistor design; the resin-bonded type, the most common type of commercial resistor; and the ceramic-bonded type, an expensive type used where close tolerance is necessary.

Resistors are formed by taking the right proportions of conductor, insulating filler and bonding material and extruding this mixture to form a long cylinder. This is then cut to the required length, connecting wires are fastened to the ends, and it is finally vitrified in an oven. The best high-frequency performance is obtained for a resistor in which the ratio of the length to the cross-sectional area is high. Because of their shape and composition, carbon composition resistors change their resistance at frequencies above about 1 MHz. Film resistors, which are made by depositing a thin layer of resistive material on an insulating former, have a high length to cross-section ratio and so are well suited to high frequency applications. In particular, metal oxide and precious metal alloy films have excellent noise properties and the resistance changes only slightly with temperature.

Wire-wound resistors are very accurate at low frequencies as they can be wound to very precise values. They are often found in secondary standard test instruments. Above 50 kHz they begin to show capacitive and inductive effects, even when they are wound in such a way as to minimize these effects.

As with all components, there is a trend towards automated manufacture and miniaturization. Eventually, discrete resistors will be largely replaced by integrated circuits and, even more important, circuit design will increasingly use inexpensive semiconductor devices instead of passive elements which are continually becoming relatively more expensive.

The colour code and preferred values

To make resistors easily recognizable a system of colour coding is used.

A typical example is shown in Fig. 1.11, the colour bands being marked A, B, C and D. The colour code is easy to

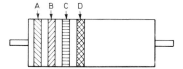

Fig. 1.11. The four band system of resistor colour coding

Table 1.2. FOUR BAND COLOUR CODE FOR RESISTORS

Colour	1st Figure 'A'	2nd Figure 'B'	Multiply- ing Value 'C'	Tolerance 'D' (%)
Silver	–	–	10^{-2}	± 10
Gold	–	–	10^{-1}	± 5
Black	–	0	1	–
Brown	1	1	10	± 1
Red	2	2	10^2	± 2
Orange	3	3	10^3	–
Yellow	4	4	10^4	–
Green	5	5	10^5	–
Blue	6	6	10^6	–
Violet	7	7	10^7	–
Grey	8	8	10^8	–
White	9	9	10^9	–
None	–	–	–	± 20

Table 1.3. PREFERRED RESISTOR VALUES AND THEIR ASSOCIATED TOLERANCES

	Tolerance			Tolerance	
$\pm 5\%$	$\pm 10\%$	$\pm 20\%$	$\pm 5\%$	$\pm 10\%$	$\pm 20\%$
1.0	1.0	1.0	3.3	3.3	3.3
1.1			3.6		
1.2	1.2		3.9	3.9	
1.3			4.3		
1.5	1.5	1.5	4.7	4.7	4.7
1.6			5.1		
1.8	1.8		5.6	5.6	
2.0			6.2		
2.2	2.2	2.2	6.8	6.8	6.8
2.4			7.5		
2.7	2.7		8.2		
3.0			9.1		

remember and with practice the reader should be able to give instantly the value of a resistor. The first colour band, A, indicates the first figure of the resistor's value, the second band, B, gives the second figure, and the third band, C, denotes the number of zeros which follows these first two figures. The fourth band, D, gives the tolerance rating of the resistor. The standard colours are listed in Table 1.2. As an example, if band A is red, band B violet and C red, the resistor's value will be 2 (red), 7 (violet) and two zeros (red) ohms, that is a 2 700 ohm resistor. With no fourth band the tolerance is 20 per cent: this means that the actual value could lie anywhere between 2 160 and 3 240 ohms. This would in practice be called a 2.7 kΩ resistor (a kilohm equals one thousand ohms).

Because of this tolerance rating, commercial resistors are usually supplied in preferred values, for it would be pointless to have resistors whose tolerance bands overlapped one another. Typical preferred values are 120 ohms, 180 ohms, 270 ohms, 1.2 kΩ, 1.8 kΩ and 2.7 kΩ, and so on, see Table 1.3. If the tolerance bands of these resistors are worked out it will be seen that they just touch one another. If it is necessary to have a resistor with a value other than the preferred ones, then it is necessary to take a selection of resistors of the nearest preferred values and measure each individually to find one of the required value. In practical circuit design such a procedure is seldom necessary, the nearest preferred value generally being good enough. Alternatively, an expensive close tolerance resistor can be used.

2 Direct and Alternating Currents

In the last chapter we talked about an electric current and how it can be produced. This type of current is called direct current (d.c.) since it flows in one direction around a circuit. A graph of the current drawn against time, that is the value of the current at given points in time, might appear as shown in Fig. 2.1. A practical use of such a graph would be in the recording of changes in amplitude of a current over a period of time due to some variation in the e.m.f. source or in the resistive behaviour of the load. But whatever the purpose, the graph shows that at intervals of time (taken on the stopwatch) the reading on the meter (indicating the current) alters by a slight amount. The graph is therefore the *waveform* of a direct supply current (or voltage).

This term, waveform, is one commonly used in electronics. Every current or voltage changes to some extent with time and thus has a waveform which in many cases is repeated over and over again.

The most common waveform is the sine waveform. This is a well-known mathematical relationship–it goes on repeating itself over a period of time with a shape as shown in Fig. 2.2.

The waveform of any energy source may be distorted by the circuitry to which it is applied and hence it is common to hear engineers say that a waveform has been *distorted*.

Alternating current

The current obtained from the mains socket of an ordinary domestic power point is usually alternating current. This type of

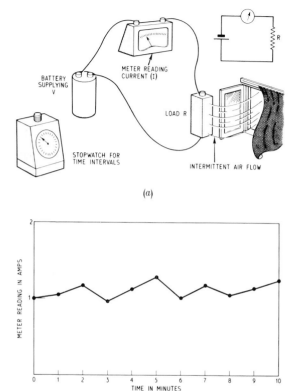

METER READING
CURRENT (I)

BATTERY
SUPPLYING
V

R

LOAD R

STOPWATCH FOR
TIME INTERVALS

INTERMITTENT AIR FLOW

(a)

(b)

Fig. 2.1. Plotting a direct current over a period of time

current is provided by motor generators which convert mechanical energy into electrical energy. The output of these generators is positive in one direction for a short period, then reduces to zero; the direction then reverses and the current gradually increases again in the opposite direction. The complete change of direction is called a cycle and is shown in Fig. 2.2. These cyclic changes or waveforms occur rapidly, the standard for mains purposes in Great Britain being fifty per second. Each complete cycle is a sinewave. Cyclic motion occurs in many natural phenomena–a common example is the children's swing which as it moves back and fourth with a varying amplitude describes what is called simple harmonic motion. If the distance moved by the swing is plotted against time a sinewave like the one shown in Fig. 2.2 (b) would result.

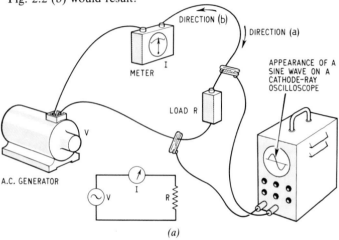

(a)

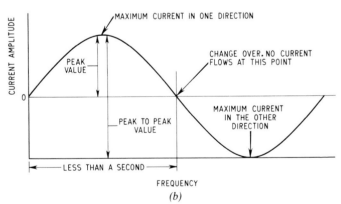

(b)

Fig. 2.2. (a) Producing an alternating current. The waveform (b) is sinusoidal

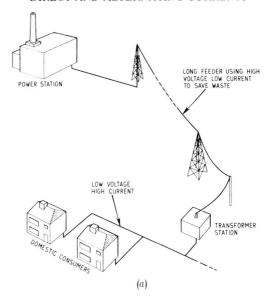

POWER STATION

LONG FEEDER USING HIGH VOLTAGE LOW CURRENT TO SAVE WASTE

LOW VOLTAGE HIGH CURRENT

TRANSFORMER STATION

DOMESTIC CONSUMERS

(a)

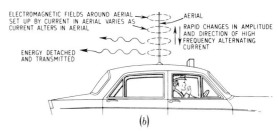

ELECTROMAGNETIC FIELDS AROUND AERIAL SET UP BY CURRENT IN AERIAL VARIES AS CURRENT ALTERS IN AERIAL

AERIAL

RAPID CHANGES IN AMPLITUDE AND DIRECTION OF HIGH FREQUENCY ALTERNATING CURRENT

ENERGY DETACHED AND TRANSMITTED

(b)

Fig. 2.3. Uses of a.c.

This sinusoidal current has a number of advantages over direct current, the most important one being that obtained by using a transformer. Currents and voltages may be changed in value and relationship with comparative ease by this means. In this way mains power can be distributed at a high voltage and low current so that the heat which would be produced by a large current flowing in the copper conductors is minimized. When the supply cable arrives near the domestic area the voltage is reduced to a safe level and current can be drawn by the user without causing great losses. This can be done only by using alternating current and transformers.

In radio communications it is necessary for radio energy to be emitted by the transmitting aerial and radiated into the ether. This is effected using alternating currents with cycle times so small that they may occupy less than one thousandth of a second. These rapid changes of current in the aerial generate electro-magnetic fields which become detached from the aerial carrying the current. In this way the radio energy is emitted and radiated. These two examples are illustrated in Fig. 2.3.

There are several important parameters to be considered in regard to sinusoidal alternating current (a.c.). These are ampli-tude, frequency and phase.

Amplitude

Amplitude may be referred to in a number of ways. Referring again to Fig. 2.2 it is possible to talk about the peak amplitude of the cycle (which may be either current in amperes or e.m.f. in volts) or the peak-to-peak value. The latter is often used in television practice, where it is called the DPA, double peak amplitude. The third and most common way to refer to the amplitude of a sinewave it to talk of its root mean square value (r.m.s.). This value is 0.7 times the peak value and is equal to the amplitude of a direct current which, flowing for the same time in the same conductor with the same resistance, would generate the same amount of heat.

Frequency

In electronics the frequency of alternating currents is an impor-tant quantity. The term is used to describe the number of complete cycles that occur in a given period of time–usually one seond. The alternating current distributed for domestic use completes one cycle fifty times in second. This is usually written as 50 hertz (Hz). The current alternating in an aerial may complete the cycle in less than a microsecond (a millionth of a second) and therefore its frequency may be several megahertz (millions of cycles per second). This is usually written as MHz.

The electronics engineer generally deals with current changes which vary from 50 Hz up to 10 000 MHz or higher. This latter frequency is so high that it becomes hard to think about, or to express, in analogous terms.

The term frequency occurs often and it is the behaviour of circuitry at various frequencies that concerns the engineer. Some examples of this are the transmission of many frequencies simultaneously over cables and the difficulty of operating valves and transistors at very high frequencies. In some spheres of electronics, for example in radio communications, it is the problem of handling these high frequencies that bothers the engineer, whereas in medical electronics, for example, it is the problem of handling very low frequencies that often causes difficulties.

In Table 2.1 the terms used for the main bands of frequencies are given. It should be noted that the term radio frequency (r.f.) which is often used is a loose term applied to electric energy alternating at any frequency above about 30 kHz.

Wavelength

In Fig. 2.4 (a) a sinewave, that is a complete cycle of alternating current, is shown to demonstrate the relationship between frequency and wavelength. So far we have discussed a current alternating in a complete circuit formed by the generator, the load and the wires connecting them. In this type of circuit it is easy to understand a current flowing first in one direction and then decreasing to zero and flowing in the other direction. The electrons are being moved back and fourth in the manner of shingle lying on the beach under the influence of the tide. However, if the circuit is made physically big enough (or the frequency made high enough) the electrons are still executing one movement when the e.m.f. changes direction. Odd effects result and the wavelength becomes an important unit in which to measure these effects.

The wavelength of a sinewave is the distance (usually measured in metres) between two points on exactly similar parts of successive cycles, for example points A and B in Fig. 2.4 (a). A practical example of wavelength occurs if a circuit is physically

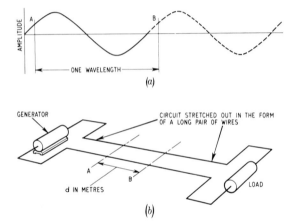

Fig. 2.4. Wavelength. (a) Two successive cycles. (b) Wavelength on a pair of wires

stretched out in the form of a long parallel pair of wires, as shown in Fig. 2.4 (b), with a generator of the sinewaves at one end and the load at the other. The voltage will be present along the wires (if they are long enough, or, if the frequency is high enough) as a series of maxima and minima. The distance between successive maxima or successive minima corresponds to a wavelength. This distance is related to frequency by the velocity of the electrical energy along the wires. Since under normal conditions the velocity is constant, the higher the frequency the smaller the wavelength. When dealing with the electrical energy in the u.h.f. and s.h.f. bands (see Table 2.1.) it

Table 2.1. FREQUENCY BANDS

Description	Frequency	Wavelength	Application
Audio frequency (A.F.) (also very low frequency V.L.F.)	30 Hz to 30 kHz	–	Sound reproduction
Radio frequency (R.F.) Low frequency (L.F.)	30 kHz to 300 kHz	10000–1000 metres	Communications, r.f. heating
Medium frequency (M.F.)	300 kHz to 3 MHz	1000–100 metres	Telemetry, control, communications
High frequency (H.F.)	3 MHz to 30 MHz	100–10 metres	Communications
Very high frequency (V.H.F.)	30 MHz to 300 MHz	10 to 1 metres	Communications and control
Ultra high frequency (U.H.F.)	300 MHz to 3000 MHz	1 to 0.1 metres	Television, communications, navigation
Super high frequeny (S.H.F.)	3000 MHz to 30000 MHz	10 centimetres to 1 centimetre	Radar, microwave devices

is usual to speak of wavelength and not frequency. The engineer talks about 3 centimetre waves and not 10 000 MHz frequency. The problems encountered at these frequencies are the concern of the microwave engineer who deals with u.h.f. and s.h.f. energy.

Phase

In Fig. 2.5, the relationship in time between an applied e.m.f. and the resultant current is shown for a sinewave. Because they are displaced they are said to be out-of-phase (the current that flows is displaced in time from the applied voltage). Each cycle is

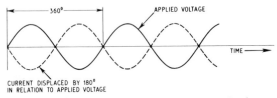

Fig. 2.5. Phase relationship between current and voltage

divided into 360° (360° is also equivalent to the distance of a wavelength) and the lag or lead of the current relative to the applied voltage is measured in degrees. In Fig. 2.5, it is 180°. Many factors can cause an alternating voltage and its current to be displaced.

The importance of phase

We can get a clearer idea of the importance of electrical phase if we consider a mechanical analogy to it. A large, heavy handcart is difficult to move: the force needed to move it has to be applied before any movement takes place and is greater than the force required to keep it in motion. This is a case of the *effect* lagging behind the *cause*. If, on the other hand, a water pump is connected to an empty tank, water will pour freely into the empty tank even before the pump applies any force and it is not until the tank begins to fill up and produce 'back pressure' that the pump has to work to continue the flow of water to fill the remainder of the tank. In this case the effect leads the cause.

In an electrical circuit the applied voltage is the cause and the current that flows is the effect, since one cannot have current flowing without a voltage being applied, although one may have a voltage applied without any current flowing.

In an electrical circuit the current can lead the voltage applied to the circuit or it can lag the voltage. If it leads it is usually because the source of voltage is applied to a capacitor (see later) which, like the empty water tank, lets the effect take place before the cause. If the current lags it may be because the voltage is applied to an inductor (see later) which is like the heavy cart in its behaviour: this time the voltage has to be applied for some definite time before the current flows.

In both cases the voltage and current do not act together and are said to be *out-of-phase* with each other. The amount by which they differ (which can be as much as a quarter of a cycle in the case of a sinusoidal wave) is referred to as the *phase difference*.

Pulse waveforms

So far we have discussed two types of current, direct and alternating. A third type has some of the characteristics of both and is called pulse-waveform current.

In its simplest form it can be produced by the circuit shown in Fig. 2.6, in which a battery is switched on and off continuously. A graph can be produced by measuring this current (either with a chart recorder or an oscilloscope) and plotting it against time. Fig. 2.7 shows this.

Looking at this graph one can see certain features of this current which are important: (1) It is direct current, that is it flows only in one direction. (2) It is intermittent, and thus shares to some extent a characteristic of alternating current. (In particular any current that changes, whether smoothly as in a.c. or rapidly as in pulsed current, can be used to transfer energy through a transformer).

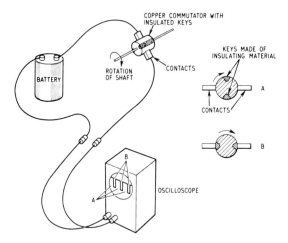

Fig. 2.6. Simple method of producing a pulse waveform by means of a rotary switch (commutator)

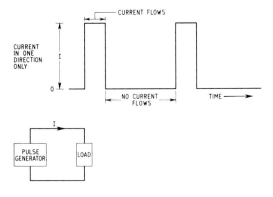

Fig. 2.7. A practical pulse waveform

Describing the pulse

To describe a pulse-waveform we do not talk about r.m.s. value as in a.c. but usually discuss the *peak value* as shown in Fig. 2.8 (a).

Whereas with a.c. we discuss frequency, when dealing with pulses it is usual to state the number of pulses per second and this is called the *pulse repetition frequency* (p.r.f.). Fig. 2.8 (b) demonstrates this for a p.r.f. of 100.

Each pulse has to be a certain length and there are intervals between pulses. The pulse length is called the 'mark' time and the intervals between pulses are termed the 'space' time. Hence the *mark/space ratio* is an important parameter when discussing the shape of pulses. Fig. 2.8 (c).

A pulse must begin at some point and then return to it for a space interval. This point is called the *reference level* and varies in different circuits. In fact the reference level of a pulse can be altered by passing the pulse through a suitable circuit. The problem of maintaining or deliberately altering a reference level is one of considerable importance to the electronics engineer.

Although we have talked about pulses of current, it is often necessary to deal with a voltage pulse which causes virtually no current to flow. This may be applied to the input of an amplifier or counting device. Some current does flow under these conditions, but is so small–for example a few microamperes–that the engineer tends to disregard it.

To describe the action of these applied pulses in any system it is necessary to know their *polarity*. As a practical example of the importance of this, a circuit designed to operate with pulses of one polarity may be able to reject a pulse of the opposite polarity. The two polarities are distinguished by calling pulses *negative-going* or *positive-going*. It can be seen from Fig. 2.8 (e), that if one terminal of the input to an apparatus is connected to the chassis of the equipment to which the pulse is applied and this is kept at zero potential to earth (that is no volts above earth) then this terminal (b) is at the zero reference level and when no pulses are present, that is during a space, the other terminal (a) is also at the zero potential. If a pulse comes along

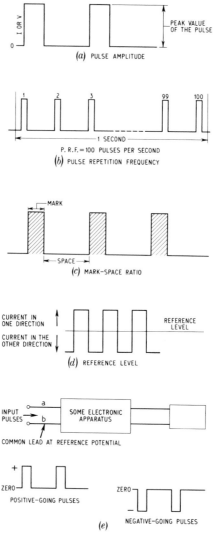

Fig. 2.8. *Pulse waveform characteristics*

and makes terminal (a) positive with respect to (b) then the pulse is said to be positive-going, whilst if it makes (a) negative with respect to (b) it is called a negative-going pulse.

In discussing circuit design, an engineer may refer to a negative-going squarewave with an equal mark/space ratio and a p.r.f. of 500 pulses per second–this being a complete description of a particular pulse waveform.

Pulse distortion

A certain amount of distortion to a pulse when it is passed through an electronic circuit is inevitable. This is caused by various factors in the circuit which delay the pulse. It is an important part of the work of the designer today to devise means of manipulating delays or time lags. Four terms are in common use to describe these delays, *delay time, rise time, storage time and fall time*. Fig. 2.9, shows at (a) a typical pulse and at (b) its shape as an output pulse after being passed through an electronic circuit. Delay time is the initial time taken for the output pulse to reach 10 per cent of its final maximum value. This is usually a very small time, for example 0.3 microseconds (μsec). More important are the terms rise time and its converse fall time, which are the times taken for the output pulse to rise and fall from 10 to 90 per cent of its total value. By ignoring the 10 per cent 'tail' portions of the pulse the engineer concerns himself only with the major change that takes place to the pulse. Storage

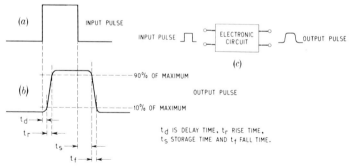

Fig. 2.9. Terms used to describe pulse distortion

time is the time delay caused by the circuit not cutting off immediately upon the cessation of the input pulse. For a modern digital integrated circuit, rise and fall (transition) times are typically 15 nanoseconds, (a nanosecond is one thousandth of a millionth of a second). Typical storage times are 10 ns.

The sinewave and the pulse

Every pulse waveform can be analysed into a number of sinewaves or, conversely, a number of sinewaves can be synthe-sized to make a series of pulses as shown, although in a rather over-simplified way, in Fig. 2.10. This makes things a bit easier because all the rules for sinewaves will also apply to pulses.

Since a number of sinewaves will combine to produce a pulse with a certain repetition frequency it is obvious that one of these frequencies must dominate and determine the repetition rate. This frequency is called the *fundamental frequency*.

Other types of waveform: the sawtooth and the spike

There are other types of waveform which the reader may encounter. Amongst them the sawtooth is perhaps the most important. This is shown in Fig. 2.11 (a). The waveform gradually increases in value as time increases: after reaching a maximum value it then falls in a very short period of time (called *flyback time*) to the zero line or reference level. The cycle is then repeated.

Another waveform commonly encountered is the spike wave-form shown in Fig. 2.11 (b). This is really a square pulse with a small mark/space ratio. When using a spike waveform the electronics engineer is concerned only with the leading edge and does not mind any distortions (such as long-tails) that may occur during the long space interval.

Alternating and pulse voltages, whether square, sawtooth, spike, etc., will all be applied to circuits and as a result of their application current will flow. How much current will flow is no longer determined by resistance alone but other factors to which we must now turn.

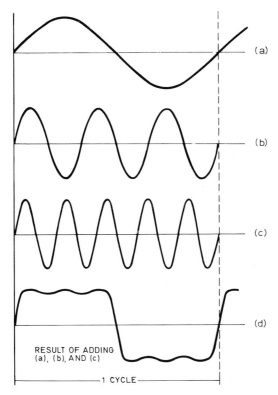

Fig. 2.10. Pulse waveform (d) derived by adding up a number of sine waves of different frequency

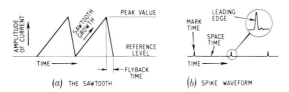

Fig. 2.11. Two other common types of waveform

The effects of a changing voltage when applied to a circuit

Just as in the case of direct currents, the amplitude of alternating and pulse currents is determined by the amplitude of the applied e.m.f. and the resistance of the conductor. However there is an additional factor which opposes the flow of alternating or pulse current and it is the result of the alternating character of such current. It is called *reactance* and its value is determined by the characteristics of the circuit through which the current flows. Every conductor possesses, to some extent, two electrical characteristics called capacitance and inductance. These characteristics can be varied by making the conductors take up certain physical shapes, e.g. by winding a wire into a coil (making an inductor) or by placing flattened conductors near to one another (the plates of a capacitor).

Both these characteristics have the effect of limiting the amplitude of the current flowing in a circuit when an alternating or pulsed voltage is applied. This effect is called reactance and its magnitude, which is measured in ohms (in exactly the same way as resistance), is determined by the amount of inductance and capacitance present in the circuit.

The value of reactance is also dependent on the frequency of the a.c.–the higher the frequency the higher is the inductive reactance and the lower the capacitive reactance. In certain configurations these reactances oppose one another and cancel out either the whole or part of each other. Reactance combines with the resistance to give *impedance*. So when dealing with alternating currents we are concerned with impedance and not just resistance.

Inductance, capacitance and resistance values depend on the physical shape of conductors, but their electrical effect, which causes impedance to current flow, does not come into play unless the current is varying. With direct currents (which are zero frequency) only resistive effects are normally operative.

Inductance

When a current flows in a length of the conductor it produces a magnetic field around the conductor as shown in Fig. 2.12. The

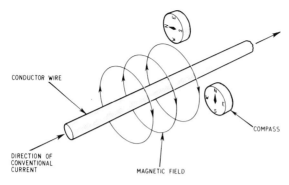

Fig. 2.12. Magnetic field round a conductor carrying a current

influence of this field can easily be detected: a compass needle held near a wire carrying a heavy current will be deflected by the field the current sets up. If the current-carrying conductor is placed adjacent to another separate conductor, as shown in Fig. 2.13 and the current is increased in amplitude, then the field surrounding the first conductor will 'swell' out and 'cut through' the second wire. In doing so it will induce an e.m.f. and hence a current in the second wire conductor.

The direction of the induced e.m.f. and its current will be such as to produce a second field which will couple with the first wire

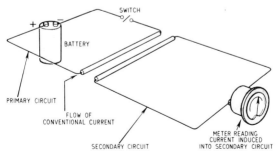

Fig. 2.13. Mutual inductance. The current flowing in the primary circuit causes a current to flow in the adjoining secondary circuit

to generate a third e.m.f. This 'reintroduced' e.m.f. is in a direction that will oppose the original current in the first wire. It is almost as though the second conductor was saying 'please leave me alone'. Its behaviour is one of reluctance to be disturbed. This is known as an inductive effect and the two conductors in this position are said to have the property of *mutual inductance*.

It should be stressed that there will be an e.m.f. induced in the secondary circuit only so long as the current is varying in the primary. Thus, if the battery in Fig. 2.13, is replaced by an alternating voltage source, an alternating voltage at the same frequency will be induced in the secondary. This is the basis of the transformer.

In the simple circuit of Fig. 2.13, the coupling between the two circuits is small and so the voltage in the secondary will be very low. A practical transformer tries to make the coupling as high as possible, and this can be done by winding the primary and secondary coils together on an iron core.

If the number of turns on the secondary is more than on the primary the voltage will be stepped up in the ratio of the number of turns. If the number of turns is smaller on the secondary the voltage will be stepped down. In an ideal transformer in which the coupling is 100 per cent there will be no power loss, but in practical transformers the coupling is below 100 per cent and there will also be losses due to eddy currents induced in the core.

This effect of inductance is an important one in electronics. Even a straight piece of conductor has some self-inductance though it is not wound in a coil or placed near another conductor. The amount is very small, but as the frequency at which the current alternates is increased the effect of even this small inductance becomes more important.

The opposition to the flow of an alternating current through an inductor increases as the frequency of the alternating current increases. The combined effect of inductance and frequency is called *inductive reactance* and is measured in ohms. It cannot be added directly to resistive ohms–the combined effect of the two when placed in series is slightly less than the arithmetic sum of both.

It will be seen that a coil carrying a direct current of constant amplitude might just as well be a straight piece of wire but, as soon as the current begins to vary, as is the case with alternating current or pulse waveforms, then the inductance begins to show itself as inductive reactance.

The effect of inductive reactance on a pulsed waveform is a little more complicated. Since this type of waveform can be considered as a large number of sinewaves of different combinations, it is clear that the inductor will present different reactances to the different frequencies and limit some more than others. The result is that a pulsed waveform current may be distorted in shape after passing through an inductor.

Units of inductance

Inductance is measured in a unit called a henry. In electronics one deals mainly in millihenries (thousandths of a henry, abbreviated mH) or microhenries (millionths of a henry, abbreviated μH). The symbol used in circuitry to denote an inductance is L.

Uses of inductors in electronics

The property of inductance is used a great deal in electronic circuits. Often it is necessary to take steps to minimize its effects, as in the case of large radio transmitting valves which when operating at high frequencies can distort the signal they are sending out if inductance is present in the wrong place. A common use of inductors is in radio receivers where, combined with capacitors, they are used to select the frequency it is desired to receive from the many available. The ability of inductors to change their reactance according to frequency means they can be used as filters to block off one frequency and allow another to pass.

Inductors are usually coils of wire wound on some insulated support called a coil former. This type of coil is suitable for many purposes but in some instances it is desirable to have a relatively high inductance with a few turns. This can be achieved by concentrating the magnetic field through the coil by placing a magnetic core in the former as shown in Fig. 2.14 (a).

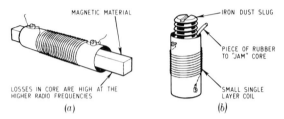

Fig. 2.14. (a) Inductor with core. (b) Typical dust core inductor

When an inductor is used at very high frequencies, unwanted currents induced in the magnetic core may reduce the 'goodness' factor, or Q of the inductor. To avoid this the following technique is used: the magnetic material is crushed into fine particles and mixed with an insulating powder and binding agent. The resultant iron dust-cores will concentrate the magnetic field but prevent unwanted currents flowing. A typical dust core and coil is shown in Fig. 2.14 (b).

Q-factor

Fig. 2.15 shows the equivalent circuit of an inductor. As can be seen there is some resistance in series with it. This resistance is that of the wire itself, and will be effectively increased by losses caused by unwanted currents flowing in the core and in any screening that may be placed around the coil. In any circuit if energy is lost in some way it is usual to regard this as being due to an imaginary resistor or, as it is called, an *equivalent resistance* in the circuit. A perfect coil would offer no resistance, having reactance only. Practical coils however offer both, for the reasons just given. The ratio of the reactance to the resistance of a coil is a measure of its 'goodness,' and is termed the Q-factor of

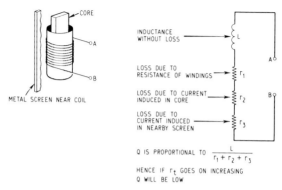

CORE

○A

○B

METAL SCREEN NEAR COIL

INDUCTANCE
WITHOUT LOSS → L

A○

LOSS DUE TO
RESISTANCE OF WINDINGS → r_1

LOSS DUE TO CURRENT
INDUCED IN CORE → r_2 B○

LOSS DUE TO
CURRENT INDUCED → r_3
IN NEARBY SCREEN

Q IS PROPORTIONAL TO $\dfrac{L}{r_1 + r_2 + r_3}$

HENCE IF r_t GOES ON INCREASING
Q WILL BE LOW

Fig. 2.15. Q-factor

the coil. The aim in designing inductors is to keep the ratio of the reactance to the resistance as high as possible. In a perfect coil, Q would be equal to infinity but in practice a coil designer may be satisfied with a Q-factor of several hundreds.

Practical inductors

Fig. 2.16 shows some of the types of inductor used in electronics. The first (a) is a typical one for radio transmitters. Its value may only be millihenries or microhenries since it will be used for frequency selection or tuning purposes at ultra high frequencies. The wire in this instance is a silver plated tube since at ultra high frequencies the current flow in a conductor tends to be concentrated in its 'skin' (hence the term *skin effect*). Each turn is carefully spaced from the other so that although the magnetic fields between turns will couple there will not be any unwanted capacitance (see later) between adjacent turns. The coil former is usually made of a ceramic material to avoid losses and so produce an inductor with a high Q.

The second inductor (b) has an iron core made of laminated iron, and a large number of turns. Its Q is very low but this does not matter for the purpose for which it is intended. It is used as a filter in power supplies and for similar applications.

Inductor (c) is a simple filter coil used to block-off radio frequency currents in some part of a circuit. Although it is shown on a former, often an engineer will wind one using a high-value resistor as a former, this combination being most suitable as a small r.f. choke.

Fig. 2.16 (d) shows a tuning coil in a screening can with an adjustable dust core. This type may be found in radio com-

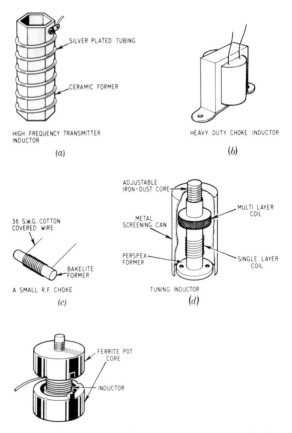

HIGH FREQUENCY TRANSMITTER INDUCTOR

SILVER PLATED TUBING

CERAMIC FORMER

(a)

HEAVY DUTY CHOKE INDUCTOR

(b)

36 S.W.G. COTTON COVERED WIRE

BAKELITE FORMER

A SMALL R.F. CHOKE

(c)

ADJUSTABLE IRON-DUST CORE

METAL SCREENING CAN

PERSPEX FORMER

MULTI LAYER COIL

SINGLE LAYER COIL

TUNING INDUCTOR

(d)

FERRITE POT CORE

INDUCTOR

(e)

Fig. 2.16. Common types of inductor

munication equipment, television and many other applications. The remaining inductor (e) is one of the latest and is used in many filter circuits, pulse forming networks and so on. The small coil is totally enclosed in a ferrite pot which can be screwed up tightly. This type of inductor gives a high inductance–many hundreds of millihenries (a high value in electronics) for only a few turns of wire, thus providing a physically small coil with a high inductance and high Q.

Capacitance

Like inductance, capacitance becomes manifest only when the amplitude and/or direction of the current in a circuit is changing.

The component called a capacitor may consist of two plates near to one another but separated by an insulator such as air. This arrangement is shown in Fig. 2.17 (a), the two plates being shown connected to a battery.

When the plates A and B (made of conductive material) are first connected to the battery the e.m.f. of the battery generates an electric field running through the connecting wires to the plates of the capacitor and across the gap between them. Under the influence of this field, electrons flow from the negative terminal of the battery. They cannot flow across the gap unless the e.m.f. is so high and the gap is so narrow that there is breakdown and a spark crosses the gap. Since they cannot cross the gap, they accumulate on plate B, charging it negatively. This negative charge repels the electrons that are in plate A. The electrons flow to the positive terminal of the battery, leaving plate A positively charged. Although part of this circuit consists of insulating material, there is a flow of electrons (i.e. an electric current) in the circuit. For as long as the current flows there is an increase of potential difference between the plates. The current flows until this p.d. equals the p.d. between the terminals of the battery, at which point it stops.

The current that flows into a capacitor is stored as an electric charge, measurable in coulombs (see Chapter 1). A capacitor with high capacitance is able to store a greater amount of charge (for a given p.d. between its plates) than one of low capacitance. Factors that determine capacitance are the area of the plates (greater area gives higher capacitance), the distance between the plates (the closer together they are, the higher the capacitance) and the nature of the insulating material, or *dielectric,* between the plates. For example, a sheet of dry paper placed between the plates increases the capacitance to double or more. The factor by

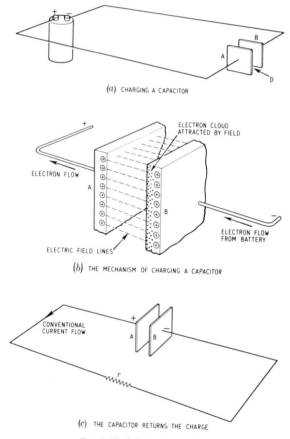

(*a*) CHARGING A CAPACITOR

ELECTRON CLOUD
ATTRACTED BY FIELD

ELECTRON FLOW

A

ELECTRIC FIELD LINES

B

ELECTRON FLOW
FROM BATTERY

(*b*) THE MECHANISM OF CHARGING A CAPACITOR

CONVENTIONAL
CURRENT FLOW

A B

(*c*) THE CAPACITOR RETURNS THE CHARGE

Fig. 2.17. Effect of capacitance

which a dielectric increases capacitance is called the *dielectric constant*. Capacitance is measured in farads. If a capacitor has capacitance 1F, it means that it stores 1 coulomb of electric charge for every volt of p.d. between its plates. The farad is very large and in practice the electronics engineer usually deals in microfarads (millionths of a farad, symbol μF), nanofarads (thousandths of a microfarad, symbol nF) or picofarads (millionths of a microfarad, symbol pF). The symbol used in diagrams for capacitance is the capital letter C.

If a capacitor is charged from a battery and then disconnected, it retains its charge for a very long time. A capacitor that has been charged while a radio set is switched on can deliver an unpleasant electric shock when touched, for a considerable period after the set has been switched off. If the capacitor is left disconnected, the charge gradually leaks away through ions present in the air or by minute leakage currents across the dielectric. If a charged capacitor is connected to a resistor, as in Fig. 2.17 (c), a current flows. The e.m.f. at which it starts to do this (i.e. to discharge) is equal to the e.m.f. which originally was used to charge it, but decreases gradually (exponentially) as the number of coulombs stored decreases.

Capacitance and alternating currents

If, instead of disconnecting the capacitor from the battery and joining the plates with a piece of copper wire, we reverse the battery terminals, as shown in Fig. 2.18 (a), we get a situation in which the capacitor will find that the existing charge it has on its plates actually aids in the process of recharging. If the battery voltage is altered in a sinusoidal manner, that is, the first e.m.f. is reduced to zero before the polarity is reversed, then the discharge current belonging to the first e.m.f. forms the charging current for the second e.m.f. This is explained in Fig. 2.18 (b).

To an alternating e.m.f. the capacitor appears to conduct electricity but the current always leads the voltage. This can be illustrated by means of the analogy shown in Fig. 2.18 (c). If a pump is pumping water first in one direction in a pipe circuit and then in another direction–that is in an alternating manner–a diaphragm placed in the pipe circuit would not stop the apparent flow of water 'current.' As the pump force (e.m.f.) drives the water back and forth the diaphragm distends and for a moment stores some water which it returns to the circuit when the pump force direction changes.

This is exactly the way in which a capacitor acts in an a.c. circuit. The degree to which this conduction occurs is a measure of the *capacitor's reactance* and again is measured in ohms like resistance and inductive reactance. Although measured in ohms,

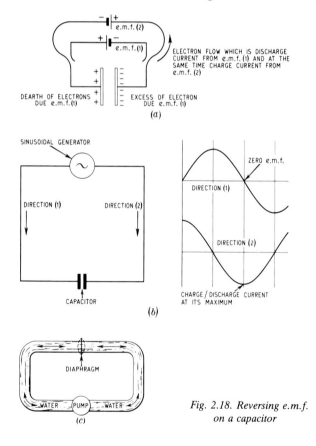

Fig. 2.18. Reversing e.m.f. on a capacitor

like inductive reactance it cannot be added directly to resistive ohms. The higher the frequency and the larger the capacitance the lower the reactance.

Capacitance and pulse waveforms

Since a pulse can be analysed into a number of sinewaves, pulses can be conducted by capacitors although the reactance will be different for the various sinewave components of the pulse. This means that the relative amplitude of the sinewave components is changed after being passed through a capacitor so that some distortion of the waveform occurs.

Practical capacitors

To explain the action of a capacitor we have assumed it to consist of two plates separated by air. In practice, however, it may be a more complicated device.

A simple air dielectric capacitor is shown in Fig. 2.19 (a). In this type the value of capacitance can be altered by varying the amount of plate enmeshed. This type of capacitor may have a capacitance of only about 200 pF and is used for 'trimming' tuning circuits.

A larger but similar type, using many plates in parallel to increase the maximum capacitance value, can be found in older radio receivers.

In the capacitors most commonly used, the plates consist of two thin sheets of metal foil. They can have a large area to give the required capacity yet are rolled with the dielectric between them to make them compact and of convenient shape. The dielectric may be paper; mica; a plastics material such as polystyrene, polycarbonate or polyester; or a ceramic material. Each type of capacitor has its own characteristics that make it suitable or unsuitable for a given application. Capacitors are made with capacitances ranging from as little as 1pF to as much as 10μF. For greater capacitance combined with reasonable size,

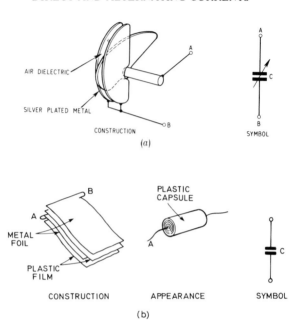

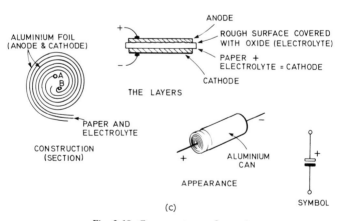

Fig. 2.19. Common types of capacitor

we use electrolytic capacitors. Fig. 2.19 (c). The aluminium electrolytic capacitor has plates of aluminium, rolled together with a sheet of paper soaked in an electrolyte (solution or paste). The main function of the paper is to separate the plates, *not* to act as dielectric. In fact, because of the electrolyte, an electric current can readily pass through the paper. The dielectric is formed by applying a p.d. to the plates. This causes a very thin film of aluminium oxide to form on the anode. This layer is non-conducting. The soaked paper is in effect part of the cathode plate, so that only a very short distance separates the plates, leading to high capacitance. Before the oxide film is formed, the anode plate has been etched to make it rough and so increase its surface area. This too acts to give high capacitance. The combination of close spacing and extra-large area makes it possible to achieve a capacitance of tens of thousands of microfarads in a capacitor of convenient dimensions. Such capacitors are convenient for storage of electric charge, but have the disadvantage that they can only work with one polarity. If a reverse e.m.f. is applied, the oxide film is destroyed. A further disadvantage in certain applications is that the coiling of the plates causes the capacitor to behave as an inductance. This reduces the effective capacitance when an alternating e.m.f. is superimposed on the steady e.m.f. applied to the capacitor, the effect depending on the frequency of the alternating e.m.f. The electrolytic capacitor behaves as if it were a tuned circuit with series resonance, like that described in the next section.

Another type of electrolytic capacitor consists of a sintered block containing particles of tantalum. For their size, these 'tantalum bead' capacitors have much greater capacitance than aluminium capacitors but are considerably more expensive and suffer the disadvantage that their breakdown voltages are relatively low.

Inductance and capacitance combined

Fig. 2.20 shows two combinations of *resonant circuit* formed from inductors and capacitors–series and parallel. Both may be used in electronics for frequency selection.

At any frequency other than the resonance frequency the first combination offers some impedance to the flow of current but, at resonance, the inductive reactance and the capacitive reactance cancel out and the only opposition to current flow is the small resistance due mainly to the inductor windings (and maybe skin effect if the frequency of operation is very high). The inductive

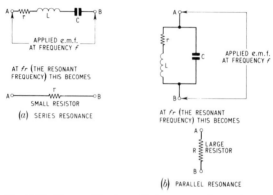

Fig. 2.20. *The combination of inductance and capacitance produces resonant 'tuned' circuits*

and capacitive reactances will cancel out at one frequency only, the one at which both reactances are numerically equal. This combination can therefore be used to allow a current at a certain selected frequency only to pass. Alternatively, the series resonant circuit may be used to short-circuit unwanted signals at the resonance frequency.

The second combination shows an inductor and a capacitor connected in parallel, forming a parallel resonance circuit. The effect of this is that at resonance current flow is minimum, the combination acting as a large equivalent resistance, but the voltage across the circuit will be maximum, as we would expect from Ohm's Law. The circuit may be used to block signal currents at the resonance frequency. Alternatively, if we are concerned with a voltage signal, it can be used to select signals at the resonance frequency. And in fact this is its most common use, acting as the load (see later) of a frequency selective amplifier.

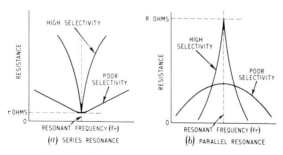

Fig. 2.21. Resonance curves of tuned circuits

The sharpness of the tuning curves shown in Fig. 2.21 (for both series and parallel combinations) is called the *selectivity* of the circuit, i.e. its ability to select one frequency from among many. How well it does this is determined by the Q of the circuit, due to the Q of the inductor which we explained earlier.

It should not, however, be thought that selectivity is the sole aim of the circuit designer. We may require a frequency selective circuit, but one that passes a wide band of frequencies. In this case the Q may be deliberately reduced by the addition of extra resistance to the circuit until the desired selectivity/bandwidth characteristic is achieved. This process is often termed *damping*.

3 Electronic Components

Every circuit consists of an arrangement of components. Some of these—resistors, capacitors and inductors—we have already discussed. Among the many other components to be found in electronic circuits are semiconductor devices, relays, inductors and transducers. One of the most important points it is necessary to know about all devices is whether they are linear or non-linear, two terms which we must know consider.

Linearity and Ohm's law

If the voltage in a circuit containing resistance, capacitance or inductance is increased the current will increase. The initial value of the current is determined by the impedance of the combined components and the value of the applied e.m.f. If the voltage is doubled then the current will double; if it is trebled the current will be trebled. This state of affairs is in accordance with Ohm's law which states that the ratio of the voltage to the current in a circuit or part of a circuit is always a constant. It is this constant that we call resistance or impedance. The components in a circuit which obeys Ohm's law are said to be *linear* elements. When certain components, for example transistors, are introduced into a circuit the simple current-voltage relationship expressed in Ohm's law no longer applies. Because of this such components are called *non-linear*. If the voltage across such a device is increased the resultant current increase does not bear a linear relationship to the voltage change. Among the simplest of non-linear devices is a semiconductor diode.

Semiconductors

As explained in Chapter 1, conduction in metals consists of a drifting of 'free' electrons, under the influence of an electric field (p.5 Fig. 1.2). In non-conducting (or insulating) materials, the electrons are tightly bound to the atoms of the crystal lattice, so no conduction can occur. We say that there are no *charge carriers*. In Fig. 3.1 we see the crystal structure of a typical example of the third class of material, the semiconductor.

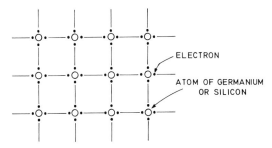

Fig. 3.1. Structure of a pure semiconductor; each atom has four electrons in its outer shell; each chemical bond (represented by a thin line) involves two electrons

Examples include silicon and germanium. This is shown in one dimension for clarity, and only the electrons of the outermost electron shell of the atom are drawn. At low temperatures these electrons are held closely by the atoms and no conduction is possible. At room temperatures and above, a number of electrons have sufficient thermal energy to allow them to escape from their positions and wander freely between the atoms. They are now able to behave as the 'free' electrons in a metal. They can act as charge carriers. If a bar of such a material is connected in a circuit such as that on p.7, conduction will occur, though the current passing will be much less than that which would pass through a bar of metal. An increase of temperature increases the number of free electrons in a semiconductor. The effect of this is that its resistance **decreases** with increasing temperature, which is an effect opposite to that observed in metals. In metals, increased temperature causes the atoms to vibrate more strongly and impede the motion of the electrons. A piece of semiconductor that it is carrying a current becomes warmer, so decreasing its resistance and **increasing** the current, which makes the semiconductor warmer still and leads to further current increase. This effect is known as **thermal runaway** and may lead to the material becoming so hot as to melt. It can be prevented by suitable circuit design, as well as by mounting semiconductors on *heat sinks*. These are plates made of heavy-gauge copper or aluminium, often with fins, that distribute the excess heat produced in devices that must carry heavy currents.

The current carried by a pure semiconductor is slight, as there are so few free electrons. These electrons are *intrinsic* charge carriers. Conduction can be increased by adding controlled amounts of other elements to the pure material. This process is known as *doping*. For example, silicon can be doped with arsenic or antimony. As shown in Fig. 3.2, the atoms of the doping

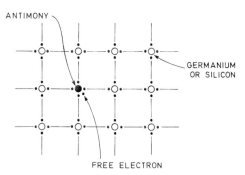

Fig. 3.2. Structure of doped n-type semiconductor; the atom of antimony has five electrons in its outer shell, so providing a spare 'free' electron

element take their place among the atoms of the semiconducting material. The doping element may be one that has an additional electron in its outer shell. If this is so each atom provides an electron that it is not firmly bound to the crystal lattice. It is 'free' to act as a charge carrier. A doped semiconductor has lower resistance than one that is not doped, because of these *extrinsic* charge carriers. A semiconductor that is doped in the way described above is called an *n-type* semiconductor, because the electrons carry negative charge.

A semiconductor may alternatively be doped with an element that has only 3 electrons present in its outer shell. The situation is illustrated in Fig. 3.3. At the site of each doping atom there is a missing electron. This vacancy is usually termed a *hole*.

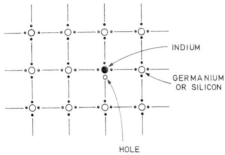

HOLE

Fig. 3.3. Structure of a doped p-type semiconductor; the atom of indium has only three electrons in its outer shell, so creating a vacancy or 'hole'

Although a hole is not a physical object but simply an absence of an electron, it can be thought of as a charge carrier. The presence of holes increases the extrinsic conduction through the material, as shown in Fig. 3.4. Although the electrons are moving for relatively short distances, from hole to hole, the *effect* is as if the holes are performing the conduction. Holes behave as positive charge carriers, so a material which is doped in this way is known as a *p-type* semiconductor. By controlling the amount and kind of doping during manufacture, semiconductors with a variety of properties can be made. Also it is easy to dope different regions of the same piece of material in different ways. Part can be p-type and part n-type. It is at the junction (called a pn *junction*) between these regions that important effects occur.

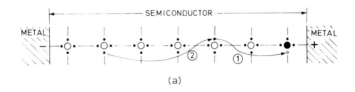

(a)

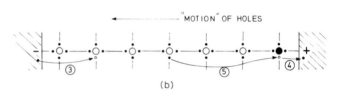

(b)

Fig. 3.4. Conduction in p-type semiconductor. (a) At 1 an electron escapes from a silicon atom and drifts in the electric field until it fills a hole at the indium atom. At 2 the hole thus created is filled by an atom escaping from an atom further along the material. The electrons drift left to right (towards +), while the vacancies or holes pass from right to left (towards −); (b) At 3, the hole created at 2 is filled by an electron from the cell to which the material is connected, having come along a metal wire. At 4, an electron escapes and passes into a metal wire and on towards the cell. The hole created is again filled by an electron, so shifting the hole toward the negative end–it acts as a positive charge carrier

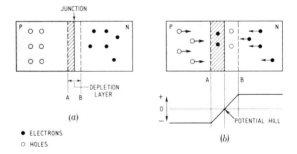

Fig. 3.5. The pn semiconductor junction

The pn junction

One way of making a pn junction is to place a small disc of metallic indium against a slice of n-type sillicon and heat them in a furnace. The indium melts and some of its atoms diffuse part of the way into the silicon. In the region with which diffusion has occurred some of the holes are immediately filled by extrinsic electrons, but the amount of indium diffused is enough to provide an excess of holes. Fig. 3.5 (a) shows the junction in diagrammatic form. Some of the excess electrons in the n-type material have enough energy to drift across the junction and fill the holes in the p-type material. On either side of the junction (from A to B) we have a region in which there are no electrons and no holes. The atoms of the material are fixed in position in the crystal lattice so are not able to drift. The loss of electrons from the n-type material leaves this region with a positive charge; the filling of holes in the p-type material produces a negative charge. A 'potential hill' is produced, as shown on Fig. 3.5 (b). This prevents the further flow of holes and electrons across the junction. It is rather as though a small cell has been 'built in' to the junction. If the semiconductor is germanium, the e.m.f. of this imaginary cell is about 0.2V. For silicon it is about 0.6V.

Applying bias to a semiconductor diode

Reverse bias.–An e.m.f. applied as shown in Fig. 3.6 (a) is said to be reverse biasing the junction. The potential hill is *increased* and the drift of holes and electrons is reduced or prevented.

   Forward bias.–If an external battery is connected to apply an e.m.f. to the junction as shown in Fig. 3.6 (b) the potential hill or gradient is *reduced* so that more electrons and holes drift across the junction and a current flows. This is called forward-biasing the pn junction.

   Thus the semiconductor diode allows current to flow in one direction only. It is unidirectional and non-linear.

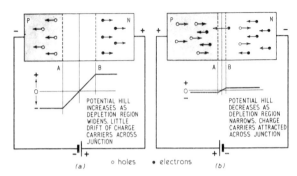

Fig. 3.6. Biasing a pn junction. (a) Reverse biasing, no current flows. (b) Forward biasing, current flows

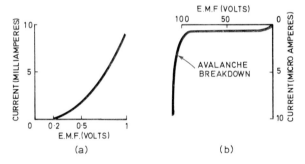

Fig. 3.7. Conduction through diode (a) forwarded biased, (b) reverse biased

These two points are illustrated in Fig. 3.7. If a forward e.m.f. is applied to a diode and gradually increased from zero volts to about 1 volt, there is no current at all until the e.m.f. of the potential hill (imaginary cell) has been more-or-less overcome. Thus a silicon diode does not conduct until the applied e.m.f. exceeds 0.6V. After that there is a gradual increase of current with increasing e.m.f. though the curve is *not* a straight line (not linear), so Ohm's Law is not obeyed. If the experiment is repeated with reverse bias (Fig. 3.7. b), there is no conduction except for a very slight leakage of a few microamperes. This is due to the presence of impurities, producing in each region a small quantity of *minority carriers* of the opposite type (eg. holes in n-type). The reverse e.m.f. may be increased to 100V or more, with gradual increase in the width of the depletion layer, but no increase in the leakage current. If a certain maximum voltage, the peak inverse voltage (p.i.v.), is exceeded the diode breaks down and is permanently damaged.

Using diodes

Semiconductor diodes are used in many applications, from computers to stabilized power supplies, and their use is steadily increasing. They have high mechanical and thermal reliability. Some typical semiconductor diodes are shown in Fig. 3.8,

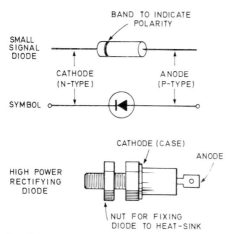

Fig. 3.8. Two typical semiconductor diodes and the circuit symbol

together with the symbol used to refer to them in circuit drawings. Diodes for power supplies are made able to carry currents of up to 6A of more, with p.i.v.s over 1000V. At the other extreme are the signal diodes used for detection in simple radio and television receivers and as elements in logic circuits. These may carry only a few tens of milliamperes but are capable of operation at very high speed.

Zener diodes

The breakdown referred to in the previous section is caused by the free electrons, accelerated by the electric field, and colliding with the silicon atoms. This knocks some of the bound electrons from position, so creating more free electrons and also holes. These electrons too are accelerated, releasing even more charge carriers. The effect is like an avalanche and causes the diode to

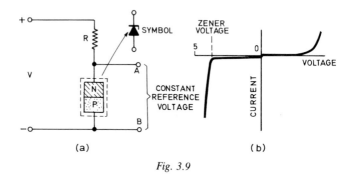

Fig. 3.9

conduct. Provided that the flow of current is limited to a safe maximum by the circuit to which the diode is connected, the diode is not destroyed. If the diode is connected as in Fig. 3.9 (a), in which V is slightly greater than the breakdown voltage, and in which current is limited by R, the p.d. across the terminals of the diode remains steady at the breakdown voltage value. If V should increase or if the current drawn by the load at A-B is decreased, the extra current is shunted through the diode. Thus the diode acts as a voltage stabilizing device. Such diodes can be made with breakdown voltages between 5V and 100V, and are generally known as zener diodes.

However, the true zener diode works in a different way from that described above. The diode is made from heavily doped silicon, with the result that the depletion layer is thin. The 'potential hill' is much 'steeper' than with an ordinary diode. Under these conditions, electrons are able to tunnel through the depletion layer when a certain reverse voltage is applied. This zener voltage can lie between 2.7V and 5V. As Fig. 3.9 (b) shows, the reverse bias characteristic of a zener diode has a sharpe 'knee' giving it good voltage stabilizing properties.

Variable capacitance diodes

Since a reverse-biased diode consists of two conducting regions with a non-conducting layer (the depletion layer) between them, it has the features of a capacitor, Fig. 3.10. If we vary the reverse

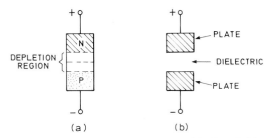

Fig. 3.10. Comparing (a) a reverse biased diode with (b)
a capacitor

voltage, the width of the depletion layer, and hence the capacitance of the diode, can be controlled. As the reverse voltage is increased the depletion zone becomes wider, so in affect producing a wider gap between the 'plates' and reducing capacitance. Special variable-capacitance diodes, or 'varicap' or 'varactor' diodes, are made with capacitances that can be varied between 2pF and 10pF. They are used in some v.h.f. radio and television circuits as tuning capacitors, with the advantage that tuning can be controlled electronically.

Tunnel diodes

It is only a few years since Esaki announced his discovery of the tunnel diode. Since the announcement a great deal of research and development work has taken place.

The Esaki or tunnel diode consists of a pn junction of heavily doped p-type material and a similar n-type material.

This heavy doping alters the normal diode characteristics and produces a negative resistance portion on the forward biased current/voltage characteristic. As the voltage across the junction is increased the current increases almost linearly until a point of saturation is reached. The characteristic then flattens and falls off so that further increases in applied voltage result in a decrease in current. This effect represents a negative resistance (since current decreases when the applied voltage increases) and makes the tunnel diode a very intersting semiconductor device. See Fig. 3.11.

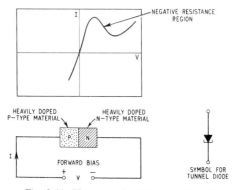

Fig. 3.11. The principle of the tunnel diode

The tunnel diode can be used in many basic circuits which could be applied to such functions as decade scaling, binary counting and other basic computing tasks. Used in conjunction with other micro-minature components, they may be made as parts of integrated solid circuits (see Chapter 5) to reduce the size of, and aid in the production of, very fast high-capacity computers of small size (Chapter 8).

Thermionic diode and triode

The thermionic diode is the original 'valve' used in the earliest electronic circuits. It was invented in 1904 by Ambrose Fleming for use as a detector in radio circuits. Like the semiconductor diode which has now replaced it, it was also widely used for rectifying alternating current. Equipment employing thermionic valves is still in use, so it is worth while knowing something about their action. The diode is shown in Fig. 3.12. When the heater

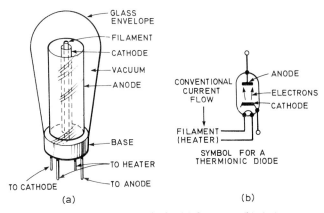

Fig. 3.12. Thermionic diode. (a) Structure. (b) Action

filament is turned on, the heating of the cathode plate (or often a cylinder surrounding the heater) causes a cloud of free electrons to be displaced from its atoms. This action is known as *thermionic emission*. If a potential difference is now applied between anode and cathode, with anode positive to cathode, the free electrons are attached toward the anode. More electrons are displaced from the cathode, and so here is a continuous flow of electrons from cathode to anode. This is a flow of electric current; the diodes conducts. If a reverse p.d. is applied there can be no conduction, for the anode is not heated and so cannot emit electrons. This property of one-way conduction is the same as that of the semiconductor diode, though the mechanism concerned is entirely different.

The triode (Fig. 3.13) is similar to the diode, but has a grid (usually of wire mesh) between the cathode and anode. By applying a varying voltage to this grid we can control the flow of electrons between cathode and anode. If the grid is at a slight positive voltage, with respect to the cathode, the grid attracts

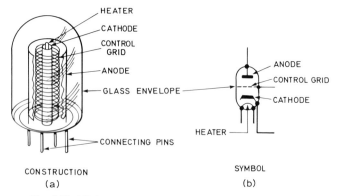

Fig. 3.13. Miniature thermionic triode construction and symbol

electron, accelerating them so that most pass through the grid and on towards the anode. The current through the triode is increased. If the grid is at a negative voltage it repels electrons and the current is decreased. By applying a small varying signal voltage to the grid we can control the relatively large current flowing through the triode. In this way the triode can act as an amplifier.

The need for amplification: the transistor

Diodes, both thermionic and semiconductor, perform many essential jobs in electric circuits. These include detecting, switching, rectifying and so on. If amplification of a signal, a basic function in electronics, is required, a more complex component than the diode is needed.

The need for amplification soon becomes apparent to the beginner in electronics. The signals with which electronics is concerned are invariably alternating or pulsed in nature and may be generated at the beginning of a system which could be a television network or telecom link or they may be produced by some action in an automation control process or even generated by the human body in a medical electronics situation. These signals may become attenuated while being passed through various electronic circuits and of course they may be very small in the first place. Before they can be used to indicate that something has happened, or to entertain someone, they must be increased in size and power.

An initial signal voltage must often be increased to perhaps a million times its initial value and the power then increased so that an output voltage is available capable of producing a reasonable current flow to operate a device such as a relay or loudspeaker.

The transistor is the semiconductor equivalent of the triode valve and has replaced it in all but a very few applications. The transistor is made by doping regions of a slice of silicon or germanium so as to produce either a region of p-type sandwiched between two regions of n-type (giving an n-p-n junction transistor, Fig. 3.14), or a region of n-type sandwiched between two regions of p-type (giving a p-n-p junction transistor). Fig. 3.14

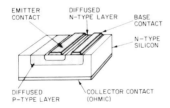

*Fig. 3.14. Diffused n-p-n silicon junction
transistor*

shows the p-type layer to be extremely thin; it is very lightly doped, so it provides relatively few holes. When connected as shown in Fig. 3.15, the base-emitter junction is equivalent to a *forward-biased* diode. Provided that the base-emitter voltage exceeds about 0.6V (see p.62) a *base current* flows between base and emitter. Let us consider the exact nature of this current. As 'conventional current' it is a flow from base to emitter, In fact, ignoring leakage there is a flow of electrons from the emitter terminal, through the n-type material to the base-emitter junction. There the electrons fill holes in the p-type material. Holes travel from the base terminal (where they are created as electrons flow out of the p-type toward the positive terminal of the cell), toward the base-emitter junction. This flow is small as there are so few holes. Typically, there is only 1 hole for every 100 electrons arriving at the base-emitter junction.

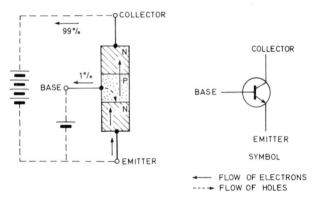

Fig. 3.15. The action of the bipolar transistor

The remaining 99 electrons, having been *accelerated* toward the junction by the field between emitter and base are able to pass straight through the *very thin* p-type layer and through the depletion layer at the *reverse-biased* base-collector junction.

They are then attracted by the much stronger emitter-collector field and pass on through the collector material and out through the collector terminal. In effect, the base-emitter potential starts the electrons off on their journey but, once they get to the base-emitter junction, *most* of them come under the influence of the emitter-collector potential. The collector current is about 100 times greater than the base current. We say there is a *gain* of 100. If the base is disconnected from the cell, no electrons arrive at

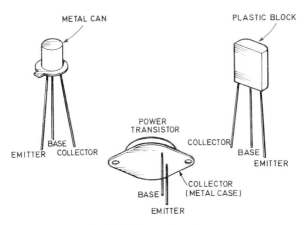

Fig. 3.16. Types of transistor

the base-emitter junction, so none can reach the collector. There is no base current and no collector current either. In this sense the transistor acts as a *switch* whereby a large (collector) current can be turned on or off by turning on or off a much smaller (base) current. If a varying current is supplied to the base, a varying number of electrons arrive at the base-emitter junction. The strength of the collector current varies accordingly. In this sense the transistor acts as a *current amplifier*, whereby the strength of a large current is controlled by variations in the strength of a much smaller current. These functions, switching and current amplifying are the two chief functions of transistors.

The transistor itself is minute but, for ease of handling, is enclosed in a case or sealed in a block of plastic (Fig. 3.16). By varying the amount of doping, the method of doping, and the

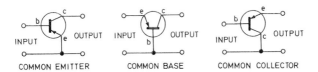

Fig. 3.17. Three basic ways of connecting a transistor

geometry of the regions, transistors of various types can be made. Some are capable of operating at high frequency (500MHz or more) for use in radio or computer circuits. Others may be designed for high gain, of 500 times or more. Power transistors, used in power-packs and high power audio amplifiers can carry currents of 30A or more.

The structure and operation of the p-n-p transistor is similar to that of the n-p-n transistor, except that polarities are reversed.

The transistor may be connected in circuit in any of the three ways shown in Fig. 3.17. There are alternative names for each arrangement. The common emitter is also known as the grounded emitter, the common base as the grounded base, and the common collector as the grounded collector or emitter follower.

The subject of transistor amplifiers is taken further in Chapter 6.

Field effect transistor

Whereas the transistor described above is called a *bipolar* transistor, because it involves *two* types of charge carrier (electrons and holes), the field effect transistor (or FET) makes use of only one type and is described as a *unipolar* transistor. The FET shown in Fig. 3.18 uses electrons since its conducting channel is of n-type material. This is the more common kind of FET, though p-channel FETs are also available. The channel allows free passage of electrons from the electrode by which they enter the channel, (the *source*) to the electrode by which they leave the channel (the *drain*). However, if the p-type region (the *gate*) is connected so that it is reverse-biased with respect to the source, a depletion region is formed in the n-type region. This happens just as in a diode (p.61). No conduction can occur in the depletion region, so it acts to make the conduction channel narrower. This increases its resistance and reduces the flow of electrons from source to drain. The greater the reverse voltage, the narrower the channel and the smaller the current.

The FET works because of the *effect* of the *field* between gate and source, and this gives it its name. Since the junction is always reverse-biased, no current ever flows across it. The only current flowing in or out of the gate is that needed to alter the potential of the gate. Since the gate region is extremely small in volume, only a minute current (a few picoamperes) flows as potential charges. In effect, the gate has exceedingly high input impedance, whereas the channel has a very low impedance. A small

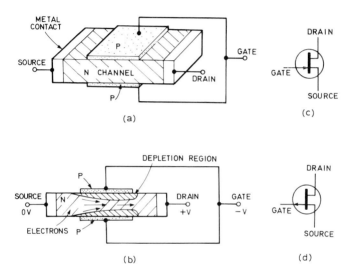

Fig. 3.18. Junction field effect transistor; (a) structure. (b) how the depletion region makes the conducting channel narrower. (c) symbol for n-channel type. (d) symbol for p-channel type

change in voltage and a negligible current can control a large current through the channel. Whereas the bipolar is a current-controlled amplifier, the FET is a voltage-controlled amplifier. FETs have many applications, particularly in amplification of voltages produced by high-impedance sources of signals, such as crystal microphones. They are also useful in voltage-measuring circuits, since they draw virtually no current and consequently do not alter the voltages that are to be measured (p.141).

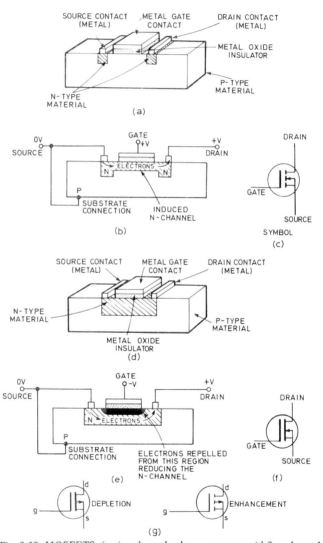

Fig. 3.19. MOSFETS; (a-c) n-channel enhancement type. (d-f) n-channel depletion type. (g) symbols for p-channel MOSFETs

Metal oxide silicon FET

This form of field effect transistor (MOSFET) is constructed in a different way from the junction field effect transistor (JFET) described in the previous section. The principle by which it operates is virtually the same, but the gate (Fig. 3.19) is completely insulated from the channel by a layer of metal oxide. This gives the MOSFET gate an extremely high input impedance. MOSFETs can be made with n-type channel or p-type channel, though the former type is more common. Either type can be constructed so as to operate in one of two modes. In the n-channel *enhancement* type no channel exists and no conduction occurs until a positive voltage is applied to the gate. Then the field induces an n-channel by repelling the holes of the p-type material. The advantage here is that there is no need to bias the gate negative of the source, as there is with JFETs. In the n-channel *depletion* type a channel exists when the gate is at 0V. If the gate voltage is reduced below zero, the channel can be reduced and completely cut off, as with JFETs. If the gate voltage is taken above zero the channel is widened and there is increased conduction. The substrate and source are shown internally connected in the figure, but sometimes they have separate terminal leads.

The operation of p-channel MOSFETS is similar, but with reversed polarities. MOSFETS of both channel types may be built up on the same silicon slice. These are *complementary* MOS transistors, often referred to as CMOS. These form the basis of a wide range of logic circuits. Their action is described in Chapter 8.

MOS transistors

One of the disadvantages of JEFTs and MOSFETs has been that they are not able to carry large currents. In recent years a new type of MOSFET, known as a v-channel MOSFET (or VMOS for short), has been produced. Fig. 3.20 shows a section through a VMOS transistor of the n-channel enhancement type. After the layers have been doped, the silicon is etched with V-shaped grooves. When a positive voltage is applied to the gate, conduction can occur down the sides of the groove. Since these are long, a wide channel is available to carry heavy current. VMOS power transistors can carry currents of 10A or more. They make it

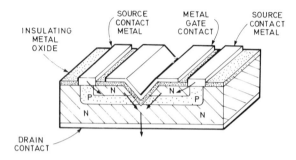

Fig. 3.20. VMOS transistor; Arrows show flow of electrons when n-channel is induced

possible to control a very large current by devices that can provide only minute currents, for example, a heavy-duty relay (p.82), or a 10W loudspeaker can be operated directly by the output of a CMOS logic gate so reducing the number of components required and cutting the costs of the circuit.

Unijunction transistor

In the UJT (Fig. 3.21) the alloying of the aluminium emitter electrode with the silicon bar produces a p-type region. If the voltage at the emitter is lower than the voltage at that part of the bar where the emitter is located, the pn junction is reverse-biased and no conduction occurs. If the voltage at the emitter is gradually raised, there comes a point (the *peak point*) at which the junction becomes forward-biased. Then conduction occurs and current flows from the emitter to base 1. Holes are swept into the base 1 region, reducing its resistance, and so allowing a

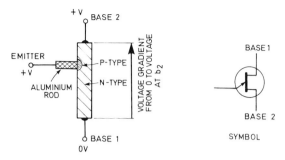

Fig. 3.21. The structure of a unijunction transistor

large current to flow. Although UJTs are not in wide use, they have certain applications in the construction of simple oscillators.

Silicon controlled switches

Members of this family of devices are variously known as silicon controlled switches (SCS), silicon controlled rectifiers (SCR), or thyristors. They are 4-layer devices (Fig. 3.22) though it is easier to understand them if we think of them as two transistors, one n-p-n and the other p-n-p, having some of their layers in common.

When first connected in circuit with zero voltage at the gate no conduction is possible. The n-p-n transistor is off, hence its collector is at a positive voltage. This means that no base current is being drawn from the p-n-p transistor, so this is also off. When a short positive pulse is applied to the gate, the n-p-n transistor is turned on; its collector voltage drops, and a base current flows from the p-n-p transistor. This is turned on and allows current to pass to the base of the p-n-p transistor. So this remains turned on, even though the pulse to the gate was exceedingly short. In other words, this is a device that can be made to conduct (in *one* direction, anode to cathode) by a brief trigger pulse at its gate. Current flows for as long as a positive voltage is applied to the anode. If this supply is disconnected, flow stops and does not start again when it is reconnected unless the device is re-triggered. Certain types of SCS have a fourth terminal wire connected to the base of the p-n-p transistor (the anode gate). A positive pulse to this wire switches off the p-n-p transistor and thus shuts off the current, even though the anode remains

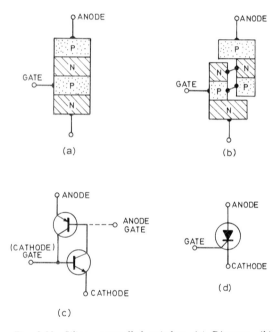

Fig. 3.22. Silicon controlled switches. (a) Diagram. (b) Represented as two connected transistors. (c) The way the transistors are joined. (d) Symbol

connected to the positive supply. These devices are used where-ever parts of the circuits must be switched on or off under electronic control. They are particularly important in power control as described in Chapter 6. The other members of this family, the triac and diac are described there.

Thermistors

A thermistor is very useful for measuring temperature. It is usually made by taking a mixture of nickel, cobalt, manganese and other oxides, forming them into a bar, a bead or a disc, and inserting two connecting leads (Fig. 3.23). The unit is then fired to give a ceramic-like appearance. As the temperature surrounding a thermistor increases so its resistance decreases (it has a

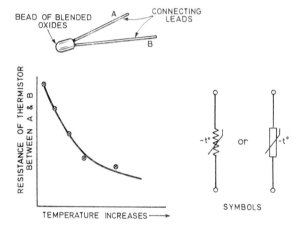

Fig. 3.23. Typical thermistor and its temperature/resistance characteristic

negative temperature coefficient). This change can be detected electronically and displayed as a temperature reading. Thermistors of small size have advantages in temperature measurement, owing to their very small heat capacity and the ease with which they can be inserted into small structures. For example, they can be used to measure temperatures inside a leaf. They are ideal for remote sensing of temperature. Thermistors are used as compensating devices to correct the behaviour of a circuit if this alters with temperature. Thermistors are also available having a *positive* temperature coefficient.

Transducers

Transducers are devices which convert mechanical energy into electrical energy or vice versa. They take many forms, for measuring velocity, acceleration, pressure, torque, distance, and so on. An example is shown in Fig. 3.24 (a). It consists of a resistance made of a coil of resistance wire with a variable tapping attached to it: any movement of this tapping by mechanical means alters, as can be seen, the amount of resistance in circuit. The change is detected by measuring the current flowing through the varying resistance in the circuit produced by a constant e.m.f. (the battery). This elementary form of transducer and variations of it are used in many applications where mechanical movement must be accurately measured.

Pressure can be measured by using a small pair of bellows whose diaphragms move with pressure change and move a variable resistance or alter the tuning of a circuit having capacitance and inductance so that the pressure changes become changes in electrical currents.

Piezo-electric material is now used for many transducers. This type of material, often lead-zirconate or barium titanate, is capable of providing an e.m.f. when it experiences any mechanical force. In the same way if electrical energy is applied to the material it will exert a mechanical force.

There are numerous applications of transducers made from piezo-electric materials–they can indicate pressure, acceleration or produce music when they are used in gramophone pickups. In ultrasonics, high frequency electrical currents are used to energize the transducer and produce correspondingly high frequency pressure vibrations which are used to do many jobs from detecting submarines to cleaning glassware. The piezo-electric transducer can be used either as a transmitter or receiver of mechanical energy.

In Fig. 3.24 (b) a piezo-electric transducer of the type used to measure acceleration in missiles is shown. Sudden change in pressure on the piezo-electric material producers an e.m.f. proportional to the acceleration.

The use of ceramics for piezo-electric devices enables transducers to be made in almost any shape with any axis of activity

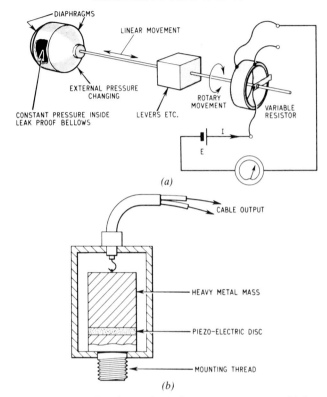

Fig. 3.24. (a) Simple transducers for measuring pressure. (b) A piezo-electric transducer designed for use as an accelerometer in missile research

according to requirements. (At London University a ceramic crystal of special type has been adapted for making delicate measurements in studying the movements of amoeba). The axis of activity of the crystal is decided and the ceramic shape fitted into a jig and immersed in special insulating oil. Across the chosen plane a polarizing field of several kilovolts per millimetre is applied for several seconds, at high temperature. The combination of high temperature and electric field reorientates the

molecules in the ceramic to create the piezo-electric properties
of the material.

The ultrasonic transducer has become of importance in remote
control or domestic equipment, in burglar detector systems
(Doppler systems) and in the ultrasonic scanner used in medicine
(Chapter 10). The ultrasonic transmitter and receeiver both
consist of ceramic material specially formed to resonate at a
fixed frequency, often 40kHz. When the transmitter transducer
is energised by an oscillating circuit operating at the resonance
frequency, it emits ultra-sound. When the receiver transducer is
made to vibrate by ultra-sound, it emits electrical signals that can
be detected electronically.

Hall-effect detector

If a bar of conducting or semiconducting material is placed in a
magnetic field as shown in Fig. 3.25, the charge carriers are
deflected. The affect is to produce an electrical field *across* the
conductor. This is known as the Hall effect. The electrical field
may be detected by a transistor connected to opposite sides of
the conductor. A Hall-effect detector incorporates the conductor
and amplifying transistors in a device which has the appearance
of an ordinary transistor. Like a transistor, it can be used for
switching current and the switching action is contact-less and
bounce-free. The switch action is usually controlled by bringing a
small permanent magnet close to the detector. The devices are
made in two forms, normally-on and normally-off, and are
switched to the other state when in a magnetic field. Such
switches have many applications in the electronic control of
machinery. If magnets are fixed to certain parts of the machine,
the positions of those parts can be monitored by Hall-effect
detectors and signals sent to the control circuits.

Relays

The relay is another important component. It is, in a sense, a
transducer since an electrical current applied to it produces a
mechanical movement. Invariably however, this movement is

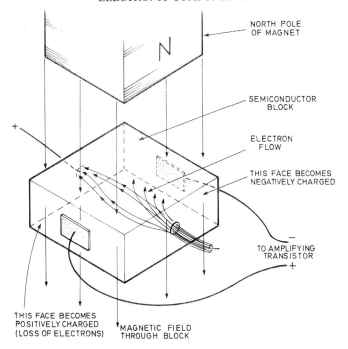

Fig. 3.25. (a) The Hall effect, showing how a magnetic field causes the electrons to be deflected

converted back immediately into an electrical change because the relay is used primarily to switch other circuits on or off.

In its simplest form the relay consists of a coil of wire wound on a piece of soft iron. When an electric current is passed through the coil the magnetic field created makes the soft iron into a temporary magnet. This 'temporary magnet' is so arranged that when it is magnetized it attracts blades of electrical contacting material in such a way as to make or break one or more other circuits, rather like an automatic switch. As the current flowing in these other circuits may be very considerably larger than the current used to operate the relay, the usefulness of the device is its ability to control large currents by small ones.

By using fast-acting relays many useful control circuits can be devised. Perhaps one of the most interesting of modern fast-acting relays is the dry-reed relay. Miniature relays of this type are now available in cases as small as those of integrated circuits. This type of relay is a fast-acting one which has almost no *contact bounce* (the effect, obviously undesirable, of the contacts bouncing on and off due to mechanical shock when the relay acts). In the dry-reed relay, two reeds of nickle-iron are placed in a small glass envelope which may be less than one inch in length. The envelope is evacuated and the leads to the reeds brought outside for external connection. The tips of the reeds are coated with gold or some other suitable material. Around the glass envelope a coil is wound and through this is passed the operating current.

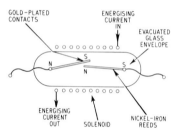

Fig. 3.26. Dry-reed relay

The current through the coil creates a magnetic field axially along the length of the glass envelope, and this field makes the reeds snap together giving a fast contact without bounce. The time taken for this to happen may be as low as a millisecond. The component is illustrated in Fig. 3.26. As the contacts are in an evacuated envelope they are free from contamination and many of the other adverse conditions affecting conventional relays. Because of the mechanical simplicity of the system, the dry-reed relay is an inherently reliable component.

4 Optoelectronic Components

The conversion of electrical energy into light energy by the filament lamp and the conversion of light energy into electrical energy by the copper-gold photoelectric cell have been known and used for many years. Since the introduction of semiconducting materials so many more new optoelectronic devices have been invented that they merit a chapter to themselves. We shall consider first those that convert light energy into electrical energy. These rely on the fact that when light shines on a semiconductor material some of the electrons are given extra energy which allows them to escape from the atom. They become free electrons and add to the number of electrons available for conduction. This effect can occur in any diode or transistor, which is why they are normally enclosed in a metal or plastic case. If the case is of glass, it is usually painted black to exclude light. Let us see how light affects the devices specially designed for the purpose.

Photodiode

This has the same structure as an ordinary diode, except that it has a clear glass case. It is used normally in the reverse-biased condition. When light falls on it, the additional electrons that are liberated join with the other minority carriers to increase the leakage current. This current can be detected and measured electronically. We can use photodiodes in circuits that measure light intentsity, or that respond to certain levels of illumination.

Certain types of photodiode are particularly sensitive to infra-red radiation. Often these are not enclosed in glass but in a plastic case opaque to visible light but transparent to infra-red. These infra-red photodiodes are particularly suited to operating in conditions where the surrounding lighting could interfere with visible light photodiodes. Applications include counting devices on factory production lines, burglar alarms and infra-red remote control of television sets.

Phototransistor

This has a similar structure to an ordinary transistor, but its case has a glass window, often in the shape of a lens, to direct light on to the transistor (Fig. 4.1). The cheaper types are encapsulated

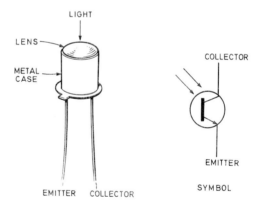

Fig. 4.1. Phototransistor of the type with no external base connection

in clear plastic. When light shines on the phototransistor the additional electrons leak through from the collector by way of the reverse-biased collector-base junction. On arriving in the base region they flow to the emitter, producing the equivalent of a base current. This is amplified by the usual transistor action, causing an increase in collector current. The phototransistor is much more sensitive than the photodiode. When used in this way there is no need to make any electrical connection to the base. In many types of phototransistor there is no base lead; the device has only collector and emitter leads.

Photovoltaic cell

One common type of photovoltaic cell is the silicon cell, often referred to as a solar cell (Fig. 4.2). A layer of n-type silicon is

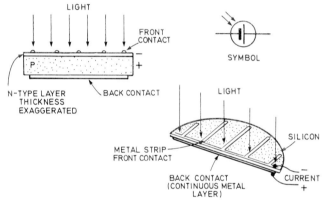

Fig. 4.2. Silicon photovoltaic cell

formed on the surface of a large slice of p-type silicon. When light falls on the depletion region, electron-hole pairs are formed and these move in the field of the imaginary cell at that region (p.61). An additional voltage thus appears, that can cause a current to flow in an external circuit.

The e.m.f. of a silicon cell in full sunlight is about 0.6V when it is not connected externally. When connected, the e.m.f. drops because of the internal resistance of the cell, but a current of several hundred milliamperes can be obtained. Solar cells can be connected in series to build solar batteries capable of an output of several watts at 30V or more. Such cells are used to power satellites and space craft as well as provide power for microwave-link transmitters and other electrical equipment located in situations remote from mains supplies. Experimental vehicles have been powered by solar cells, including an aeroplane that successfully flew using power from batteries of solar cells arranged on its wings. At present, solar cells do not have high efficiency. They convert only 15% of incident light energy to electrical energy. There is much scope for increasing their efficiency. Research to achieve this, as well as reduce their cost, is actively under way. Low-cost cells based on copper indium, selenide and cadmium sulphide promise to have high efficiencies.

Photoconductive cells

These devices, sometimes known as *light dependent resistors* (LDR) come into a slightly different category to the photocells above. They do not produce an e.m.f. when illuminated, but simply show decreased resistance. They consist of a disc of semiconductor material (usually cadmium sulphide) on which are printed two interlacing conductors (Fig. 4.3). The whole is

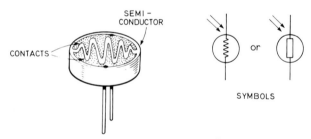

Fig. 4.3. Photoconductive cell

then encapsuled in clear plastic. In darkness the resistance of the device is high, usually about 10MΩ; when light falls on the device, electron-hole pairs are generated. Its resistance therefore falls in proportion to the intensity of light. In bright sunlight the resistance falls as low as 100Ω. Photoconductive cells have many applications for measuring light levels. Modern photographic exposure meters usually make use of this. They can operate in a wide range of light intensities, but their chief

disadvantage is that they are slow to respond to changes in the
level of illumination. A photodiode or phototransistor can
change state in less than a microsecond, but a typical photoconductive cell takes about 75 milliseconds to respond to a fall of
light intensity and several hundred milliseconds to respond to a
rise. This makes photoconductive cells unsuitable for rapid
counting operations or for reading punched tape or optical sound
tracks.

Light emitting diodes

We now turn to the other class of optoelectronic device, the
photo-emissive devices, which convert electrical energy into light
energy. By far the most popular of these is the light emitting
diode, or LED. In the commonly used sort the semi-conducting

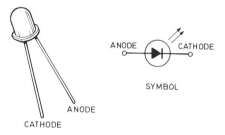

Fig. 4.4. Light emitting diode

material is gallium phosphide or gallium arsenide. When the
diode is forward-biassed it emits light. A big advantage of LEDs
is that they operate on very low voltages, typically about 2V.
This makes them ideal for use in connection with battery-
powered equipment especially equipment involving integrated
logic circuits (p. 111).

LEDs are available in a range of colours: red, yellow, orange
and green. They are also made to emit infra-red radiation but no
visible light. This is the type used in the transmitter of infra-red
remote-control systems. One type of LED has two diodes in one
capsule, arranged with opposite polarity, so that the LED shines

red when the current flows in one direction and green when it flows in the other.

The light emitting diode is enclosed in a plastic capsule which may be assembled to form a 'bar' indicator. The amount of light emitted by a LED is small, but is more than adequate for an indicator lamp or an information display.

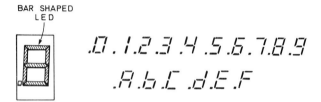

Fig. 4.5. LED 7-segment display and the characters it can produce

The LED is the basis of the 7-segment display so frequently used in watches, clocks, calculators, and cash registers (Fig. 4.5). It can display any digit from 0 to 9 with or without the decimal point. It can also display the letters A to F, and a few others, and is used for this purpose to display hexadecimal numbers in microprocessor systems.

Gas discharge lamps

The neon lamp is still in use as an indicating lamp, though rapidly being replaced by the LED. It needs an e.m.f. of at least 70V, a fact which restricts its use to mains-powered equipment. Indicator lamps are often made with built-in series resistors so that they can be run directly from the a.c. mains supply: this type is popular as a 'power-on' indicator. Small tubes filled with neon or other gases are used in the 7-segment configuration to produce numerical displays. These too require relatively high voltage, but they provide a very bright display, so are frequently used in mains-powered equipment, such as cash-registers and video recorders where e.m.f.s of 100V or more can easily be provided.

Lasers

A laser is a device in which a large number of atoms are excited in such a way that they all emit light radiation in phase with each other. It is able to produce a beam of light of extremely high intensity, with high energy content. In industry and in scientific research, lasers can be used to produce a high concentrations of energy in a small area, to generate exceedingly high temperatures and pressures. Laser beams have travelled from Earth to the Moon and back. The laser is not an electronic device so a description of how it works is beyond the scope of this book. They have applications associated with electronics in the field of telecommunications. Miniature lasers are used to generate modulated beams of high-intensity light that are directed into one end of an optical fibre. The fibre is a narrow strand of transparent plastic perhaps only a millimetre or less in diameter, but a kilometre or more long. The fibre is coated with another transparent plastic of higher refractive index, so that total internal reflection occurs. This means that a very high percentage of the light from the laser arrives at the far end of the fibre. There it is detected by photo-sensitive electronic devices. By this means, optical signals may be sent along the fibre, just as easily as electrical signals can be sent along a telephone wire. Unlike electrical signals in a wire, laser-produced light signals are uneffected by electromagnetic disturbances, so are free from interference. Also, a very large number of fibres may be twisted together, without need for insulation between them. This means that a 'cable' of given diameter can carry many more simultaneous transmissions than a conventional telephone cable of equal diameter. Fibre-optics communications links are installed in several sections of our telephone system and their use is expected to spread.

Liquid crystal displays

These operate on an entirely different principle from the devices described in the previous three sections, for they do not emit light, but simply act to effect any light that passes through them.

The liquid crystal display consists of two glass plates with a thin layer of the liquid crystal material sandwiched between them (Fig. 4.6). The inner surfaces of the glass are printed with a transparent film of conductive material such as indium oxide or tin oxide to form electrodes. The liquid crystal is an organic

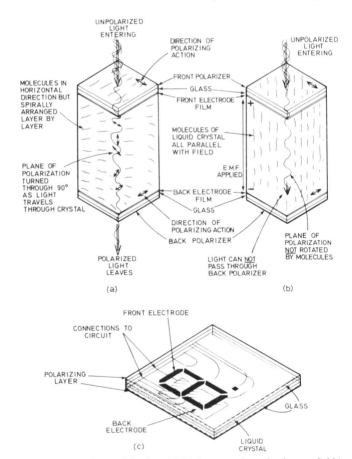

Fig. 4.6. Liquid crystal display; (a) Light passes through when no field is applied. (b) Light is stopped when a field is applied. (c) 7-segment LCD

substance that is liquid but, unlike most liquids, has crystalline properties. Its molecules are long and narrow in shape and the naturally-occurring forces between them tend to make them lie together in a regular patten, hence their regular crystal-like properties. The most important result of these properties is that they are able to rotate the direction of polarization of light as it passes through the crystal. As Fig. 4.6 (a) shows, the light may be rotated a quarter of a turn by a film of the correct thickness. If polarizing films are placed on either side of the cell as shown, light passing in will be rotated by 90° and pass out at the other side. The crystal appears transparent. If light is reflected back again from a mirror, it can make the return journey too.

If an e.m.f. is applied across the electrodes, the electric field alters the arrangement of the molecules (Fig. 4.6 (b)). Now there is no polarization; light polarized at the front surface is *not* rotated, so cannot emerge at the back. The region of liquid between the electrodes appears *black*. When the e.m.f. is removed, the molecules almost instantly return to their former crystalline arrangement and the liquid becomes transparent once again.

In a liquid crystal display (LCD) the back electrode usually covers the whole surface, and a pattern of separate electrodes is printed on the front surface. This can take the form of a 7-segment display or any other symbols or words required. By applying an e.m.f. between the back electrodes and some of the front electrodes, selected regions of the crystal can be made to appear black. In this way, the device may be made to display figures, symbols or words, as required.

The illumination for an LCD comes from external sources –natural lighting or a built-in filament lamp. The display may be viewed by reflected light, with a small mirror behind it to increase contrast, or by transmitted light with an illuminated panel behind the LCD. Whereas an LED display is not bright enough to be seen in full sunlight, an LCD is as clear as ever. Since the LCD does not provide its own light energy it requires only enough energy to provide the electric fields. The current required is only a few microamperes, compared with tens of milliamperes for an LED display. This is why the LCD has become so popular for battery-powered portable equipment such as watches and calculators. It can be provide a continuous display yet the small batteries have a long life.

Opto-coupled devices

These consist of a light-emitting device, usually a LED, and a light-sensitive device, such as a phototransistor sealed in the same capsule (Fig. 4.7). The phototransistor receives light from

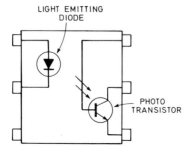

LIGHT EMITTING DIODE

PHOTO TRANSISTOR

Fig. 4.7. LED and phototransistor in a 6-pin case, forming an opto-coupled device

the LED, and the case is opaque so as to exclude light from outside. Such devices are used when signals must be sent from one circuit to another but is not suitable for the circuits to be electrically connected. For example, the LED may be controlled by a logic circuit operating at low voltages, while the transistor may be part of another circuit in which high voltages are present. The two sides of the device are electrically insulated to withstand voltage differences of several thousand volts. Signals transmitted may be simple 'on-off' digital signals or continuously varying analogue signals. In the former case, the device has the function of a simple relay, though its action is considerably faster.

Opto-coupled thyristors and triacs (p.123) are used to switch alternating currents at main voltages by means of low-voltage control circuits.

5 Integrated Circuits

Before the 1950s when transistors came into common usage, all electronic processes such as signal forming, detection, gating, and amplification were performed using thermionic valves. These devices were large and consumed high powers to perform even the most delicate tasks because of the need to heat a filament and use high anode voltages. This also meant that bulky power supplies had to be used, and other circuit elements such as capacitors and inductors had to be large to withstand the high circuit voltages.

Miniature valves and components were developed and mounted on printed wiring or printed circuit boards to simplify manufacture. However, these components were still large compared with those in common use today.

The real impetus towards component miniaturization was given by the invention of the transistor. When it became widely accepted, its small size reduced equipment volume considerably, and its low operating voltages meant that other components could be made smaller, further reducing equipment volume. To take full advantage of this small size, circuits were redesigned to eliminate transformers and large air-spaced variable capacitors. Printed circuit boards soon became used almost universally for wiring transistor circuits, only now complete circuits could be realized on a printed board smaller than a playing card.

In the early 1950s, junction transistors cost several pounds to buy and could only work at relatively low frequencies (up to about 1 MHz). They were used mainly in expensive equipment such as hearing aids where the small size and the low cost was of prime importance. By the 1960s, the price had fallen many times to much less than a pound sterling so that transistors could be used widely in commercial radio sets.

Small and inexpensive though transistors then were, compared with valves, scientists were designing electronic devices of ever increasing complexity–such as digital computers–in which huge numbers of transistors would be required. Quite apart from the problems of connecting together 100 000 or more transistors by hand, the size and cost of such systems would have been enormous. This made it necessary to reduce components' sizes still further, but this time by a factor of a hundred times or more.

The breakthrough came in the late 1960s, when better materials and new semiconductor techniques made it possible to construct complete circuits–including capacitors and resistors –on a single chip of silicon no bigger than the head of a nail. These were known as *integrated circuits*. Individual components on the chip no longer had to be connected by hand. Instead, all connections could be made at the same time by evaporating thin conducting films on to the chip to interconnect components in a similar way to printed boards, but on a microscopic scale.

Early integrated circuits packed 50 to 100 components on to a single chip, but this number increased rapidly until thousands of components were being manufactured on a single chip. The development of the metal oxide silicon (MOS) transistor enabled even higher component densities to be realized.

Component miniaturization brought with it many advantages in addition to small size. Before the 1950s, a large number of different materials were needed in an electronic circuit: tungsten, nickel, and glass for valves; carbon for resistors; ceramics and aluminium for capacitors; copper for inductors; and so on. Also, many discrete components do not lend themselves to modern methods of automated manufacture and so are relatively expensive. On the other hand, integrated circuits use only one basic material–silicon–together with minute quantities of other materials, and are well suited to modern manufacturing techniques, thus keeping costs low. In addition, they are more reliable and have a longer life than discrete components, permit more complex circuits to be developed, and have enabled new specialized logic techniques to be developed for computers, control systems, and so on.

Most important of all, the microscopic size of the individual transistors increased their operating frequency so that it became possible to use them in microwave applications in the gigahertz region, and for very high speed switching in computers. An additional bonus was that the very short paths between components no longer limited the operating speed of a transistor circuit, whereas in discrete component circuits the speed was often limited by the presence of several feet of connecting wire.

Circuit functions

Electronics can be divided functionally into two parts. In the first we seek to create shapes–sinewaves, square waves, spikes, and so on. These shaped signals have to occur at particular points in time and with definite relationships to each other. Also, we must be able to detect these shapes, recognize them, modify them, and alter their timing. In other circuits we simply open or close a gate in response to the correct input signal.

Power does not play a part in any of these operations. As long as the signal is there (and is not swamped by noise) it can be manipulated by very low power devices. Semiconductor devices are now used universally for such applications.

In the second part, when the shaping and timing of the signal has been finalized, it may be necessary to amplify the signal, perhaps for transmission over a radio link or to operate electromechanical equipment such as a tape reader. Here the power transistors are required.

Printed boards

Conventional circuits using valves, resistors, capacitors, and inductors used to be, and occasionally still are, connected together by hand using copper wire. However, in the late 1940s, as components became smaller, printed boards came into use.

A printed board is made from a flat piece of plastic insulator on to which is laid a suitable pattern of thin copper conductor strips that connect the individual components which are soldered on to the board in predetermined positions. In some cases a small amount of hand wiring with copper wire may still be necessary as it is not always possible to design the conductor layout to completely eliminate conventional wiring. A good design should reduce hand wiring to a minimum. In complex circuits, designing a board may take a long time, and it is becoming more common to design the boards using computers.

When transistors came into general use in the 1950s, printed boards became the usual method of component interconnection, many circuits being built on printed boards only the size of a playing card, or smaller. The next step for more complex circuits was to use multilayer printed boards that connected components in three dimensions, thus further miniaturizing circuit assemblies.

Components can be soldered on to the board by hand, or the process can be automated using techniques such as flow soldering, that solder on all components automatically.

Solid-state integrated circuit components

As transistors became smaller and production techniques more sophisticated, a stage was reached at which normal discrete passive components were much larger than the active element. It became necessary therefore to find new ways of fabricating resistors and capacitors, ways that would be compatible with semiconductor techniques. If this were possible, then entire circuits could be constructed on a single substrate to give a truly integrated circuit.

Solid-state resistors

Fig. 5.1 (a) shows a resistor formed by diffusion techniques on a semiconductor substrate such as silicon.

A semiconductor has, as its name suggests, a finite resistance. Using the formula

$$R = \frac{\rho l}{tw} \quad \text{ohms}$$

where ρ is the resistivity of the semiconductor used, w is the

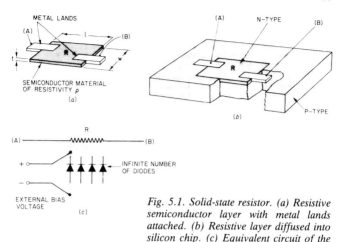

Fig. 5.1. Solid-state resistor. (a) Resistive semiconductor layer with metal lands attached. (b) Resistive layer diffused into silicon chip. (c) Equivalent circuit of the isolation technique

width of resistor, t is the thickness of resistor, and l is the length of resistor, it is possible to calculate the size of a piece of semiconductor material with the desired resistance. The material can be n-type or p-type, it does not matter which at this stage. Two metal tabs or lands are now attached to the material and we have a normal resistor of minute proportions.

The next problem is the siting of this resistor on the substrate which might also be used as the substrate for the transistor that will use this in its circuit. We want to connect the resistor to the transistor in a special way to give the circuit configuration we desire. Therefore, the general siting of the resistor on the substrate must not be allowed to interfere with this. In fact the resistor must be electrically isolated from the substrate that supports it. This is achieved, as shown Fig. 5.1 (b), by using a resistor of say n-type material with a substrate of p-type material. This forms a diode junction which can be reverse biased by an external voltage to isolate the resistor from the substrate material in which it rests.

The equivalent circuit for this isolating technique is shown in Fig. 5.1 (c).

Solid-state capacitors

Normal capacitors, see Fig. 5.2 (a), consist of two conducting plates separated by a dielectric of air or some insulating material. When a voltage is applied, current flows into the plates and distorts the dielectric which stores this electrical energy. If a growing voltage is applied to the plates, a large current flows intially which reduces as the voltage increases. This phenomenon causes a phase shift between the voltage and current in which the

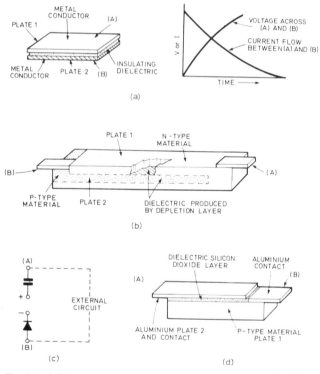

Fig. 5.2. Solid-state capacitor. (a) Conventional capacitor. (b) Diffused solid-state capacitor. (c) Equivalent circuit . (d) Solid-state capacitor using silicon dioxide mask as the dielectric

current leads the voltage. Both these effects can be produced by a semiconductor diode which is reverse biased.

Fig. 5.2 (b) shows such a device. The depletion layer between the n-type and p-type materials behaves as a dielectric, while the p-type and n-type materials on either side act as the conducting plates. It is reverse biased to prevent the junction behaving, as it naturally would, as a diode. Fig. 5.2 (c) shows the equivalent circuit.

Fig. 5.2 (d) shows a second type of capacitor that uses the aluminium contact as one plate, silicon dioxide (which is used to mask the silicon during diffusion and is therefore nearly always present) as the dielectric, and p-type silicon as the other plate.

Solid-state integrated circuits

As it is virtually impossible to fabricate a practical inductor in semiconductor material, circuits using all solid-state components must use only diodes, transistors, and capacitors. This limitation is not so harsh as it may seem. Oscillators and amplifiers can be designed for a wide range of frequencies, and pulse-forming and selective circuits are also possible. Fig. 5.3 (b) shows the circuit diagram of a basic amplifier that can be constructed in integrated circuit form.

In an integrated circuit all the components are fabricated on one chip, usually silicon. The transistors, diodes, and resistors are fairly easy to arrange but capacitors need relatively large areas.

Once the components have been formed by masking and diffusion processes, they are connected to each other by an aluminium pattern deposited on top of the silicon chip.

Today, a chip containing a number of transistors, diodes, capacitors and resistors, can be produced on a silicon chip only 0.15 mm square.

Fig. 5.3 (a) shows the final appearance of the chip before it is packaged. Its size may be less than 0.15 mm in length. The capacitor is formed by aluminium contact A, silicon dioxide dielectric, and the n-type material below. Contact at B is made by another aluminium conductor that connects the capacitor to the base electrode of the n-p-n transistor. The emitter comes at C via a third contact. The collector connects to D, a fourth conducting strip, which connects the collector to the resistor formed by the n-type block of silicon. The other side of the resistor is taken to the contact E.

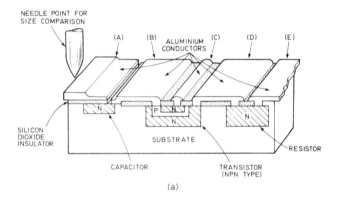

(a)

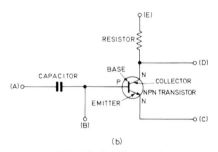

(b)

Fig. 5.3. Simple amplifier

The whole integrated circuit lies on a substrate of p-type silicon. This substrate must always be kept negative to the n-type material so as to create a reverse diode effect and electrically isolate the three components from the substrate that supports them.

Thin and thick-film components

Thin films are widely used in microelectronics as a method of producing stable close-tolerance resistors and close-tolerance capacitors. Another important application is the use of conducting films to interconnect the individual components in integrated circuits of thin-film circuits, rather like a printed board is used to connect discrete components but on a much smaller scale.

It is also possible to make transistors and diodes using thin film, but because of the high quality of planar semiconductor components, techniques for thin-film devices have been widely developed.

The basis of the construction of thin-film components is the laying down of a number of thin films (i.e. films less than about 2×10^{-5} mm thick) one after the other on to an insulating substrate. The most important techniques for depositing the films are evaporation and sputtering. The substrate is suitably masked before a film is laid down so that the correct film pattern is obtained. By using several different masks and a number of successive depositions, complete-thin film circuits can be laid down on a single substrate.

Thin-film resistors

These are the simplest of the thin-film components to produce. A thin film of resistive material is laid down between two highly conducting tabs or 'lands' on an insulating substrate such as glass or ceramic, as shown in Figure 5.4 (a). The length, width, and thickness of the film must be carefully chosen and carefully controlled during manufacture, to produce a close-tolerance resistor of the required value. The most common materials used for the resistive films are metals, metal alloys, amd metal/dielectric mixtures.

Higher-value resistors may be obtained by meandering the resistive film as shown in Fig. 5.4 (b). Alternatively, a film of higher resistivity can be used.

Where very close tolerance resistors are required for special applications, these can be produced by trimming a resistor, that is, by removing a small part of the film to increase the resistance slightly.

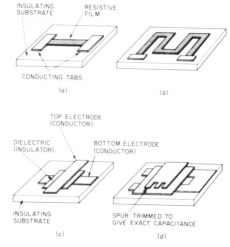

Fig. 5.4. Thin-film components. (a) Simple thin-film resistor. (b) Meandered resistor. (c) Simple thin-film capacitor. (d) Trimming a thin-film capacitor

Thin-film capacitors

There is a close resemblance between discrete capacitors using a thin sheet of dielectric sandwiched between sheets of metal foil, and thin-film capacitors in which a thin dielectric film is sandwiched between metal film electrodes. Construction of a thin-film capacitor is basically simple, as shown in Fig. 5.4 (c), but requires very precise techniques to give close tolerances. Three films must be deposited in turn on the substrate. First, the conducting bottom electrode, second, the dielectric film, and finally, the top electrode.

If a capacitor is required with a closer tolerance than the deposition technique allows, the capacitor can be made with spurs which are then trimmed to give the precise capacitance required.

By using different dielectrics and varying the dielectric film thickness, a wide range of capacitances can be realized. However, the dielectric cannot be made too thin or breakdown may occur. Common dielectric materials are silicon monoxide, aluminium oxide, and tantalum pentoxide.

When building resonant ot 'tuned' circuits (see p.54) the conductor may be replaced by an amplifier circuit. This is the obvious thing to do when trying to make an *integrated* resonant circuit since the action of an inductor depends on its physical size and it is impossible to miniaturize it. At the same time, it is a very simple matter to fabricate the amplifier in an i.c. This combination is called an *active* filter, for an amplifier is an active device. The selectivity or 'cut-off' of such a filter is much sharper than that of the passive filters described on pp.54-55. In active filters (Fig. 5.5 (a)) the resonance frequency depends on the value of the resistor and capacitor. At audio frequencies it

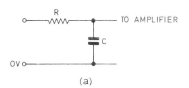

(a)

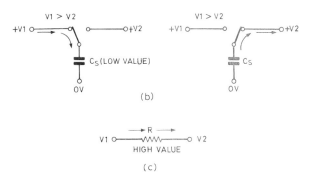

(b)

(c)

Fig. 5.5. (a) Simple RC resonant circuit. (b) Action of a switched capacitor Cs. (c) Equivalent resistor

happens that either R or C (or both) must be fairly high. It is difficult to make an integrated capacitor of sufficiently high capacitance: there is a limit to how thin the dielectric can be, and to have plates of large area is obviously ruled out. The capacitor *must* be small. If C is to be small, R must be large. This means a *long* resistor, which will take up too much space also. The solution to this dilemma is to use a low-value (i.e. small size) capacitor and to substitute a *switched capacitor* for R. Fig. 55 (b) shows how it works. (Note that we do not really employ a mechanical switch as shown here; switching is done by transistors.) With the switch to A, C charges to V_1. Then with the switch to B it discharges to V_2 (assuming V_2 is lower than V_1). Each time the switch changes, a quantity of charge passes across from A to B. The amount of charge passed each time depends on the value of C_s. If C_s has very low capacitance, the amount is very small. The effect is the same as having a high-value resistor between A and B. By using a switched capacitor we can build the filter circuit from two small-valued (= small-sized) capacitors. These together with the amplifier and switching circuits, are integrated on a single chip.

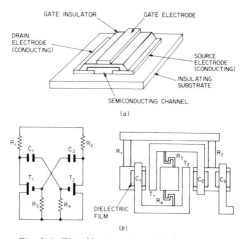

Fig. 5.6. Thin-film transistor. (a) Cross-section.
(b) Multivibrator built using thin-film technique

Thin-film transistors

The construction of a thin-film transistor (or TFT) is shown in Fig. 5.6 (a). Is is similar in operation to the MOST described previously as it has an insulated gate electrode and is controlled by varying the field between this gate and the active semiconductor film. However, there are some differences in detail. The TFT is fabricated on an insulating base, and the source and drain contacts with the semiconductor film are ohmic, whereas in the MOST they form pn junctions with the substrate.

TFTs and thin-film resistors and capacitors can be fabricated on a single substrate, forming a complete thin-film circuit. A simple multivibrator circuit using all three components is shown in Fig. 5.6 (b), together with the circuit diagram.

Thick-film components

These are very similar to thin-film components but, as the name implies, they use thicker films–usually from about 1 to 5×10^{-2} mm thick.

The tolerances of these components are poorer than for thin films, but fabrication is simpler and more easily automated. The films are basically printing ink with different materials added to give the required properties. For example, metals are added for resistors and conductor patterns, and metal oxides are added for capacitor dielectrics.

Film patterns are laid down on the insulating substrate by a highly accurate silk screen printing process, and the films are then fired to bond them to the substrate.

Thick films are used almost entirely for resistors, capacitors, and component interconnection.

Hybrid circuits

In many integrated circuits which include both active and passive elements, the accuracy of diffused resistors and capacitors is

insufficient to meet the circuit requirements. It is therefore
necessary to use other types of resistor and capacitor, either
discrete components or film circuits being preferred because of

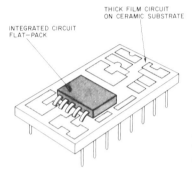

THICK FILM CIRCUIT
ON CERAMIC SUBSTRATE

INTEGRATED CIRCUIT
FLAT—PACK

*Fig. 5.7. Hybrid circuit consist-
ing of an integrated circuit flat
pack mounted on a thick-film
circuit*

their small size. Circuits which are made by combining integrated and film techniques are known as hybrid circuits.

Fig. 5.7 shows a typical hybrid circuit in which a flat-pack
integrated circuit is mounted on and connected to a thick-film
circuit.

Large-scale production

The designing of an integrated circuit requires a considerable
amount of investment of time and resources and it is therefore
very expensive. But once the design is complete and has been
proved by practical tests, the production of large numbers of the
circuit is very inexpensive. By large-scale production and
marketing the high development costs can quickly be recovered
and the circuit can be sold very cheaply yet profitably. It is this
ability to produce integrated circuits cheaply in large numbers
that has brought high-quality radio and television sets, tape-
recorders, television games, electronic watches, programmable
washing machines, calculators and home computers into the lives
of so many of the population.

The first stage in production is the growing of a large cylindrical crystal of the substrate material, usually silicon, which is then sliced thinly. Each slice is then put through the various stages of fabricating the integrated circuit, as described above. The interesting point about this operation is that hundreds of individual circuits can be built simultaneously on the same chip. Instead of using a single mask to produce just one circuit, the mask can be multiplied in rows or columns to produce hundreds of copies on one slice. These can all be processed by one series of operations at virtually no greater cost than would be used for one circuit. When the processing is complete, the slice is placed in an automatic testing machine, which steps along the slice regularly, pressing test-probes to the terminal lands of each circuit. The circuit is put through a series of tests in a fraction of a second. If it fails any test, it is automatically marked with a spot of paint.

When testing is complete, the slice is cut into separate 'chips,' each bearing one circuit. Those that have been marked as defective are rejected. Microscopic flaws in the silicon slice or faults in the processing may cause a high proportion of the circuits to be rejected, perhaps as many as 50%. In spite of this and because the quantities of raw materials required are so trival, a complex circuit can be manufactured for very low cost. During the decade of the seventies, when the large-scale manufacture of integrated circuits first got into full swing, the price of a complicated integrated circuit fell to what had been the price of a single transistor ten years before.

Practical integrated circuits

Although the integrated circuit chip may be only a millimetre or so square it must be connected to the external circuit through terminals. These must be reasonably large and spaced so that they can fit into sockets or into holes drilled in the printed circuit board. A number of standard cans and packages have been developed (Fig. 5.8), of which the most commonly used one is the dual-in-line or d.i.l. package. This and the flat package are fabricated either in metal, or ceramic material or in plastic. The plastic package is the cheapest and most commonly used type.

Within the package there is a framework of conductors leading from the terminal pins. The circuit chip is mounted in the centre. Terminal lands on the chip are connected to the terminal pins by thin gold wires soldered at either end. Special machines are used to perform the soldering quickly yet precisely. Although d.i.l. packages may vary in their number of terminals from 6 pins to 40 pins, the majority have 14 pins or 16 pins and look monotonously alike. Yet the variety of type numbers printed on the packages testifies to the wide range of types available. The main types of i.c. fall under the following broad headings.

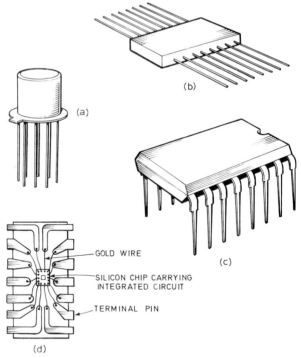

(b)

(a)

(c)

GOLD WIRE

SILICON CHIP CARRYING
INTEGRATED CIRCUIT

TERMINAL PIN

(d)

Fig. 5.8. Integrated circuit packages. (a) Metal can. (b) Flat-pack. (c) DIL package. (d) DIL package with upper half removed showing connections to the circuit

(1) Linear i.c.s. These include amplifiers of many kinds. Some are specially designed for audio applications and many of these operate at high power. Those of the highest power, often over 200W, are provided with heat sinks to dissipate excess heat. For use in stereophonic radios and audio systems, i.c.s. often have two identical amplifiers on the same chip, one for each channel. Radio-frequency amplifiers are also available in integrated form. One of the most versatile of amplifiers is the operational amplifier, or *op-amp*. It is a high-gain amplifier with special features that make it useful in many applications, including audio amplification, and the detection and measurement of very small e.m.fs. High-precision versions of operational amplifiers are used in analogue computers. Practical applications of an operational amplifier are described on p.134.

(2) Logic i.c.s. The action of these is described in Chapter 8. They are also known as *digital i.c.s.* The early logic i.c.s provided only the simplest of functions such as the elementary logical operations NAND and NOR and required only a dozen or two of integrated components on one chip. These i.c.s were followed by circuits or greater complexity, such as counters, storage registers, small memories and arithmetic devices that could add, say, two 4-digit binary numbers. This level of complexity is known as medium-scale integration (MSI). It was quickly followed by large scale integration (LSI) and very large-scale integration (VLSI) in which whole logical systems are fabricated on a single chip. The pocket calculator and the digital clock or watch usually contains but a single i.c. to perform all these operations. The heart of the microcomputer is a single microprocessor chip. Microprocessor i.c.s (or MPUs) can be manufactured so cheaply that it is economical to incorporate them in automatic machinery such as washing machines, sewing machines, industrial robots, automobiles and aircraft control systems even though any one particular application does not make full use of the microprocessor's abilities. A microprocessor can be controlled by a special logical program, incorporated in a special memory i.c., so that any one type of MPU can be made to perform one of a variety of special tasks.

(3) Special function i.c.s. It is impossible to describe the wide range of i.c.s that come under this heading. We have timer i.c.s that, with a few external components, can control time periods from a few microseconds to several days in length. We have the phase-locked loop i.c., that can be tuned to pick out and lock on to a signal of a particular frequency from a mass of other signals. There is the radio i.c., looking just like a transistor that, with a tuning capacitor plus 2 or 3 resistors, gives a complete radio set. There are the remote-control i.c.s that automatically generate a coded stream of pulses: when this code is received at the television set another i.c. decodes the message and controls the changing of wavebands, the brightness of the picture or the volume of the sound. There seems to be no limit to the possibilities of integrated circuits.

Even more exciting is the technique of fabricating an enormous number of circuits on a substrate and then using sophisticated test procedures to check out and connect up only a proportion of these circuits in a chosen configuration to produce the particular type of circuit required. This approaches very closely to biological systems like the brain where there are masses of redundant cells, and in which only those pathways and cells needed for a special purpose are used.

Looking even further into the future, it could be possible for circuits to detect a fault within themselves and repair the system by finding a new interconnection path via previously unused components. This is rather like teaching someone whose brain has been damaged to use new cell links in order to restore speech or movement.

6 Basic Electronic Circuits

The components which we have been discussing in the previous chapters can be arranged in various ways in electronic circuits in order to fulfil a number of basic functions. In all electronic operations we are concerned with a signal, which may be an alternating current oscillating at a very high frequency, a series of pulses or spikes, a waveform which varies in amplitude or a direct current which is fluctuating slowly. In some electronic equipment this signal may be fed in from an outside source and require, for example, detection and measurement; in other instances we may have to generate the signal, apply it to some outside process, and then later detect it again in order to see what has happened to it in the meantime. There are a number of basic circuits that play a part in all or a few of the many possible applications of electronics. In nearly every application, for example, we need circuits which will *amplify* a weak signal; in may applications we need various circuits which generate various different types of signals—oscillators to provide high frequency a.c. signals and timebase generators to provide pulses and spikes. Also there are a number of basic circuits using diodes or simple resistance-capacitance networks which we use in order to shape or otherwise modify the signals. In this chapter what can be considered as the important basic circuits are outlined.

Switching circuits

Before we consider switching circuits in detail there is one simple circuit element that must be explained. This is the *potential*

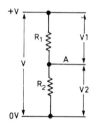

Fig. 6.1. Potential divider

divider. It consists of two or more resistances wired in series. In Fig. 6.1 the total p.d. across the two resistances is V. By Ohm's Law the current flowing through them must be

$$I = \frac{V}{R_1 + R_2}$$

Taking each resistance separately we can now calculate the p.d. across each. For the lower resistance the p.d. is $V_2 = IR_2$. This equation can be arranged to give

$$I = \frac{V_2}{R_2}$$

Since the current, I, referred to for R_2 is the same current as referred to in the equation for the *whole* resistance chain, we can say that

$$I = \frac{V}{R_1 + R_2} = \frac{V_2}{R_2}$$

or

$$V_2 = \left(\frac{R_2}{R_1 + R_2}\right) \times V$$

If we increase R_2 while keeping V and R_1 constant, V_2 increases. The equations assume that no current is being drawn from point A. If we connect point A to another circuit, this could effect the potential at A. In practical circuits, we ensure that the current I is at least 10 times the maximum current that will be drawn from A, so that potentials at A will not be unduly affected.

In the simplest switch circuit (Fig. 6.2) the base current to the transistor is controlled by a switch, S. When this is closed, current flows through R (which limits the current to a safe value) to the base of TR. As explained on p.70, this causes a collector

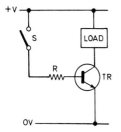

Fig. 6.2. An n-p-n transistor as a switch

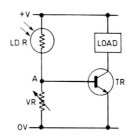

Fig. 6.3. Light triggered transistor as a switch

current to flow. The transistor is turned *on*. Since the load is in the collector circuit, a current now flows through the load. The load may be a lamp or the coil of a relay. If the switch is opened, no base current flows and the transistor is turned *off*. Then there is no collector current and the lamp goes out or the relay is de-energised. We might replace S by a pair of contacts dipping into water. The current that could flow through such a 'switch' is enough to switch on a transistor. Such a circuit is used as a water-level detector.

In Fig. 6.3 we see how a transistor may be switched by a potential divider. Here the resistances are a light-dependent resistor, LDR, and a variable resistor, VR. First of all, VR is adjusted so that the potential at A is *not quite* enough to provide sufficient base current to turn TR on. For a silicon transistor, the potential at point A might be set at, say, 0.4V. If the illumination reaching LDR now increases, the resistance of LDR decreases. This has the effect of increasing the potential at A. Now the transistor is switched on. A collector current flows and the load is energised. The setting of VR determines at what level of light intensity the switching action occurs. If desired LDR and R can be interchanged so that TR is switched on as light levels are decreased. Such a circuit could switch on room lighting at dusk each day. The LDR can also be replaced by other devices, such as a thermistor, photodiode or phototransistor. With a thermistor in circuit we have the basis of a thermostat.

The circuit of Fig. 6.3 can be improved by using a second transistor. In Fig. 6.4 the variable resistor VR is acting as a potential divider. Its two sections above and below the point of

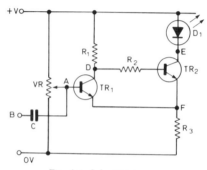

Fig. 6.4. Schmitt trigger

contact of the wiper act as the two separate resistances of Fig. 6.1. As the wiper is turned from one end of VR to the other, the potential at A can be set to *any* value between 0 and +V. Let us assume that it is set so that TR_1 receives just sufficient base current to turn it on. If TR_1 is off, it passes no collector current: it behaves as a high-value resistor. If we think of the chain R_1-TR_1-R_3 as a potential divider, the section TR_1-R_3 has very high resistance. This means that the potential at point D must be high, almost equal to +V. A current is flowing through R_1 and R_2 so TR_2 is on. This means that a collector current flows through TR_2; the LED, D_1, is lit. If the potential at A is increased slightly, for example by a positive pulse from B passing through capacitor C, TR_1 is turned on. This makes the potential at D fall; the values of resistors are chosen so that this fall reduces current to TR_2 sufficiently to turn it off. D_1 is extinguised. When the pulse is over, the potential at A returns to its former value, but this does *not* mean that TR_1 is turned off again. Let us see why this is so. Originally, when TR_2 was *on*, there was a current flowing through R_3; therefore there must have been a p.d. between its ends. Point F was a little above 0V. The potential at A had been adjusted so that TR_1 was less than 0.6V. *Now*, with TR_2 *off*, no current flows in R_3 and both its ends are at *0V*. The potential difference between A and F is now bigger than it was originally, big enough to produce a base current that will keep TR_1 switched on. So TR_1 stays on and TR_2 stays off. The result is that this circuit can be triggered to change state by a single positive pulse. To change it back to its original state it needs a *negative* pulse at B; this will reduce the p.d. between A and F to less than 0.6V, turn off TR_1 and so put TR_2 on again. This circuit is stable in two states, so it is known as a *bistable* circuit. There are several other bistable circuits (see next section), but this kind is called a *Schmitt trigger*. As shown in Fig. 6.4, it will give a square-wave output at E when a sinewave or similar output is applied at B. The trigger can also be operated by the potential divider of Fig. 6.3. As light slowly decreases D_1 goes out at a certain light intensity. If the light intensity then increases slightly D_1 does not come on again. It comes on again only when light intensity decreases by an appreciable amount. It is the property of Schmitt trigger circuits that, having been made to change state by a small change of input potential, they will not change back again until there has been a much larger change of input in the reverse direction.

Bistable circuits

The *monostable multivibrator* (Fig. 6.5) has many features in common with the Schmitt trigger. A high pulse at A is transmitted through C_1 and acts to turn TR_1 on. This causes a drop in potential at B, for the same reasons as given for the Schmitt trigger. This low pulse is transmitted through C_2, reducing the current to TR_2 and turning it off. The output of the circuit at F goes high. Let us follow what happens at the plates of capacitor C_2. To start with its plate X was at $+V$ volts (or almost) since TR_1 was off. Its other plate Y was at the relatively low voltage of point D. In the potential divider R_5-TR_2-R_4, TR_2 has a low voltage drop of 0.6V, the forward bias drop of its base-emitter junction; R_4 has low resistance, but R_5 has high resistance. So there is a substantial p.d. between X and Y. Let us say X is at 9 volts and Y at 2 volts, a p.d. of 7 volts. As soon as the circuit is triggered, the voltage at B drops suddenly to about 2 volts.

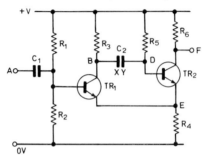

Fig. 6.5. Monostable multivibrator

Under this sudden change the capacitor cannot rapidly lose its charge. *It maintains the p.d. between its plates.* X has dropped from 9 to 2 (a fall of 7V) so Y also drops by 7V, from 2V to −5V. With a potential of −5V at D, there is 14V p.d. across R_5 and current begins to flow to Y but not to the base of TR_2. Since TR_2 is now off, potential at E has dropped to 0V; this ensures that TR_1 stays on, even though the original triggering pulse at A is finished. As current flows through R_3, the potential at plate Y gradually rises from −5V. The rate at which it rises depends on the values of R_5 and C_2. If these are large it may take several minutes or tens of minutes for the potential at Y to rise to 0.6V. As soon as it reaches 0.6V, TR_2 begins to conduct, output voltage at F begins to fall. The potential at E begins to rise, so TR_1 is turned off. The circuit returns to its original state with TR_1 off, TR_2 on, and F at low voltage. The action of this circuit is to produce a *long* output pulse from F whenever it is triggered by a *brief* pulse at A. By choosing suitable values for R_5 and C_2 this output pulse can be made as long as we choose, within reason. The monostable circuit is a useful one for producing delays and for timing purposes.

In the *astable multivibrator* (Fig. 6.6) we again have two transistors, but cross-coupled by capacitors. At any one instant one transistor is on and the other is off. Let us begin with the circuit just having changed from one state to the other. TR_1 has just turned off and TR_2 has just turned on. Output A has just gone low. The drop in potential at A affects C_2. The potential at plate X falls sharply, taking the potential of plate Y below zero, just as described for the monostable circuit. TR_1 is held firmly off while current flows through R_3, gradually charging Y. The

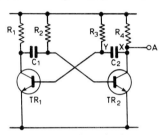

Fig. 6.6. Astable multivibrator

potential at Y (and hence at the base of TR_1) rises until it reaches 0.6V. Then TR_1 begins to turn on, the potential at its collector and C_1 starts to fall, TR_2 is turned off and the potential at A rises from low to high. The circuit is then in its other state. C_1 charges through R_2 and, when the potential at the base of TR_2 reaches 0.6V, the circuit reverts to its original state. This circuit automatically alternates between its two states at a rate depending on the time taken to charge its capacitors. Thus it oscillates at a frequency depending on the values of C_1, C_2, R_2, and R_3. Circuits of this kind are widely used as oscillators in audio-generators and alarm circuits, for flashing warning lights and many other applications.

Before leaving the bistable circuits we should look at the simplest of them all, the flip-flop, Fig. 6.7. Let us begin with TR_1 off, TR_2 on and output D low. The circuit is stable in this state. If a brief low pulse is applied to input A, nothing happens, for the base of TR_1 is already low. If the pulse is applied to B, TR_2 is momentarily turned off. The potential at D rises, current can now flow through R_3, turning TR_1 on. The potential at E therefore falls, no current can flow to the base of TR_2 so it *stays* off, even though the low pulse is finished. The brief pulse at A has made the circuit 'flip' from one state to another. Further low pulses at B have no effect. Only a low pulse at A can make it 'flop' back to its original state. By measuring the output of a flip-flop, we can tell which one of its inputs was *the last* to receive

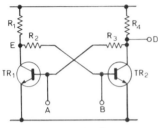

Fig. 6.7. Flip-flop

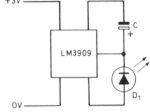

Fig. 6.8. Astable multivibrator based on an integrated circuit

a pulse. The circuit can 'remember' which input received the last pulse. Flip-flops have very wide applications as memory units. They were used in the earliest computers to store data and are still used in that way when only a few items of information need to be stored. For the storage of the masses of information used by a modern computer, we now use integrated circuits that operate on very much the same principle.

Bistable circuits appear in various forms as basic units in a host of electronic devices. Because they are widely used they are all available as integrated circuits (p.106 and 163). Usually the integrated forms make use of many more than two transistors, giving them such useful characteristics as fast rise and fall times, low current consumption, improved temperature stability and highly reliable triggering at precise voltage levels. As an example of circuit integration consider the LM3909, Fig. 6.8. The i.c. includes all the components for an a stable multivibrator except the timing capacitor, of which only one is needed. Instead of eight components, we need to wire up only two which saves time, space and cost. This i.c. can be used as the basic oscillator in a wide range of other circuits.

POWER SUPPLY CIRCUITS

One of the chief functions of most power supply circuits is to convert the 240V mains supply, alternating at 50Hz, to a direct current supply at the much lower voltages required by modern solid-state equipment. The reduction of voltage is usually achieved by a step-down transformer. In a few instances, where the current required is very small, it may be more convenient to use a dropper resistor. The transformer current is still an alternating current and must then be *rectified*. The simplest way to do this is by using a single diode, as in Fig. 6.9. The diode

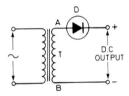

Fig. 6.9. Single-diode rectifier

conducts on each half cycle when end A of the secondary coil is positive to end B. In the other half cycle it does not conduct, for it is reverse-biased. Thus we obtain half-wave d.c. from this circuit, as illustrated by the second waveform of Fig. 6.10. This arrangement, though simple and cheap, is inefficient and the intermittent supply of current is unsuitable for many types of circuit. By arranging four diodes in a *bridge* (Fig. 6.11), there is conduction during both halves of the cycle. When A is positive to C the route of the current is A-B-load-D-C. On the other half-cycle the route is C-B-load-D-A. Bridge rectifiers can be purchased as a single unit, containing four diodes connected as shown. The output of a bridge rectifier appears as the *full-wave* d.c. of Fig. 6.10. This fluctuating voltage too may be unsuitable for many circuits but it can be smoothed by connecting a capacitor across the supply. In Fig. 6.11, C is an electrolytic capacitor with capacitance of several thousands of microfarads. During the peak of each half-cycle the capacitor receives a boost of charge from the rectifier. The charge passes out to the

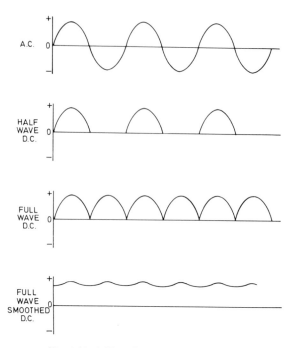

Fig. 6.10. A.C. and rectified d.c. waveforms

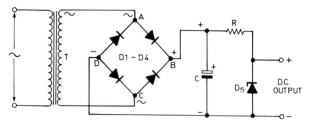

Fig. 6.11. Bridge rectifier, with smoothing capacitor and zener
diode for voltage regulation

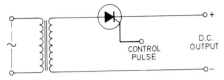

Fig. 6.12. Using a silicon controlled rectifier or thyristor

powered circuit at a steady rate so there is a *slight* fall of voltage after each peak. If the capacitance is sufficiently large, voltage will have fallen by only a small amount before the capacitor is recharged to the next peak. The waveform of smoothed d.c. is a more-or-less steady voltage with slight *ripple* at 100Hz.

Fig. 6.12 shows a rectifying circuit employing a thyristor (p.77). This does not conduct at all unless a positive trigger pulse is applied to its gate. When triggered, it behaves as the diode of Fig. 6.9. In thyristor circuits, the pulse is automatically applied once during each positive half cycle. The pulse-generating circuit

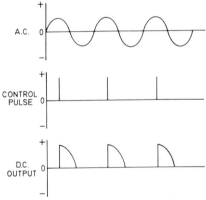

Fig. 6.13. A.C. and thyristor controlled d.c.

allows the pulse to be generated during any desired stage of the half-cycle. Fig. 6.13 shows what happens. Here the pulse is applied just as the a.c. voltage reaches its peak. The thyristor begins to conduct and the output is as shown as the lower graph.

However as the a.c. voltage falls to zero and goes negative the thyristor becomes non-conducting. From then on, it does not conduct again until it is triggered *half-way through the next half-cycle*. Current flows for only a quarter of cycle. By altering the timing of the trigger pulse we can arrange for conduction to begin earlier or later in the half cycle. In this way the average current flowing to the output can vary between the maximum (equivalent to half-wave d.c.) and zero. Circuits of this type are in common use for controlling the brightness of room light and the speed of motors on equipment such as electric drills.

The thyristor shares the disadvantage of the diode, that it can not conduct for more than half the time. The *triac* Fig. 6.14 is a

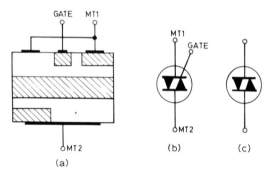

Fig. 6.14. Triac. (a) Structure. (b) Symbol. (c) Diac symbol

solution to this problem. Comparison with Fig. 6.22, p.135, shows it to resemble two thyristors connected in opposite sense, but sharing the gate electrode. It is to be triggered in *each* half cycle so the power obtainable can be varied between zero and that of full-wave d.c. A simple circuit for this function is shown in Fig. 6.15. The alternating potentials in the circuit result in rapidly changing potentials between one plate of capacitor C and the variable resistor VR. By alternating the setting of VR, the timing of the charging and discharging of C may be altered. This part of the circuit is a resistance-capacitance phase shifter. It allows trigger pulses to be generated at any required stage during the a.c. cycle. Triggering is effected by a device known as a *diac*. This can be thought of as two diodes connected in reverse sense, as its symbol implies. The diodes are ones in which avalanche breakdown (p.64) occurs at relatively low voltages. When the voltage across the diac exceeds this voltage breakdown occurs, generating a pulse that triggers the triac.

The regulation of voltage by a zener-diode was illustrated in Fig. 6.11. This is a simple technique and is satisfactory provided that the current drawn from the circuit does not exceed the calculated range. There are regulating circuits that use transistors as variable resistors so that the voltage remains stable under

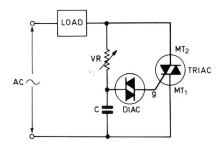

Fig. 6.15. Full-wave power control circuit

a wide range of conditions. Such circuits may also incorporate *current limiting* features. If the load develops short circuits or if the current drawn by the load exceeds a safe amount for any other reason, the supply voltage is sharply reduced and little current is allowed to flow. There is not enough space to go into the details of such circuits here. Nowadays there are numerous types of integrated circuit for voltage regulation and current limiting. These can deliver an output at a prescribed or controlled voltage though the supply voltage may fluctuate widely and the current drawn by the load may range from a few microamperes to several amperes.

SIMPLE AMPLIFIERS

One of the most common operations necessary in electronics is amplification. Signal voltages and currents are often very small initially and, as they pass through equipment, losses which produce *attenuation* of the signal my occur. They must therefore be amplified before they can be useful. The output of a radio communications transmitter, for example, may be very high but several hundred miles away it may be capable of producing only a few millivolts of signal in a radio receiving aerial. This is too small to operate a loudspeaker or recording chart so that amplification is necessary before use can be made of it. In medical electronics the small signals present in the human body are measured and presented as a record to the clinician for his assessment: these signals are usually very small, often only microvolts, and they must be magnified considerably before a full understanding of their significance is possible.

Gain of an amplifier

Engineers talk about the *gain* of an amplifier and this is generally expressed in decibels (abbreviation dB). The decibel is a unit used to describe the ratio of two quantities. Sounds are also usually measured dB. A road drill, for example, is said to be as noisy as 90 dB. This means that the ratio of the noise of the drill to some other noise, for example, a whisper, when expressed in decibels, equals 90. In amplifiers it is the ratio of the output voltage to the input voltage that is given in dB as the gain (provided that the input and output impedances are the same). Fig. 6.16 shows a box representing an amplifier. The input voltage is 1 millivolt and the output is 10 volts. This means that the amplifier has amplified 10 000 times or, expressed in decibels, it has a gain of 80 dB.

The useful thing about decibels is that they can be added directly together. For example, if an amplifier with a gain of 50 dB has its output connected to the input of an amplifier with a gain of 100 dB, then the total gain is 150 dB.

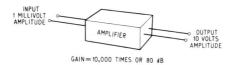

Fig. 6.16. Amplifier considered as a 'black box'

In Table 6.1, a range of decibel values against actual gain is shown. It can be seen from the table that the dB figures increase more slowly than the other ones. For example, an amplification of 10 times equals 20 dB and yet one of a 1000 times equals only 60 dB. It may be seen that the dB method of stating gain is a

Table 6.1

Vout/Vin	Gain in dB	Vout/Vin	Gain in dB
1	0	80	38
2	6	100	40
4	12	200	46
8	18	400	52
10	20	800	58
20	26	1000	60
40	32	10 000	80
		100 000	100

misleading one but in practice it enables the designer to keep a sense of proportion. Amplification itself is not enough, it is good signal-to-noise ratio that counts (in all electronic equipment there is a certain amount of background interference, called *noise,* caused by random electron movements, etc.).

Transistor characteristics

Conduction through a transistor does not obey Ohm's Law. The value of the collector current depends, among other things, on the strength of the base current. The collector-emitter path can be thought of as a variable resistor. The changes in the resistance of the base-emitter junction as different base currents are applied are transferred to the collector-emitter path and appear there as changes of resistance. The transistor gets its name from this property. It was first known as a *'transfer resistor,'* later shortened to 'transistor.' The properties of a given transistor may be measured and plotted as a series of graphs. If constant base current (I_b) is applied and the collector-emitter p.d. steadily increased, the collector current (I_C) rapidly increases to maximum value. Further increase of collector-emitter p.d. causes no appreciable increase of I_C, until breakdown voltages are reached. The curves slope only imperceptibly upwards over a wide range of p.d. For a given base current the collector current is virtually constant. Yet varying I_B has a marked effect. Fig.

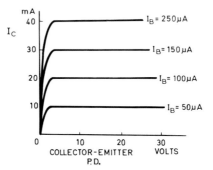

Fig. 6.17. Characteristic of silicon junction transistor with gain of 200

6.17 shows that I_C is directly proportional to I_B. In the example chosen, I_C is 10 mA when I_B is $50\mu A$. This is a current gain of 200 times. If I_B is doubled, to $100\mu A$, I_C becomes doubled to 20 mA. If I_B is trebled, I_C is trebled. Above a certain value, further increases of I_B do not produce corresponding increases of I_C. The transistor is entering the stage of being *saturated* and I_C is approaching its maximum. Before this step is reached there is a linear relationship between I_C and I_B. This is the range in which a transistor should operate so as to amplify without distortion.

Common emitter amplification

Most amplifiers employing junction transistors use the common emitter or grounded emitter connection (see p.72). In Fig. 6.18 the base of the transistor is connected to a potential divider, R_1-R_2. The values of the resistors are such that sufficient base

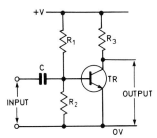

Fig. 6.18. Simple common-emitter amplifier

current flows to bring the transistor into a conducting state. A steady collector current flows through the transistor and R_3. Flow through R_3 *does* obey Ohm's Law. So there is a p.d. between its ends, proportional to the current. The output voltage, measured from the 0V line, equals +V *minus* the p.d. across R_3. A steady potential at the input has no effect for C simply maintains a steady p.d. between its plates. It neither gains nor loses charge. If an alternating current, such as a speech or music signal from a crystal or magnetic microphone is applied to the input, charges in current cause changes of potential at the capacitor. These are transmitted across C to the plate connected to the R_1/R_2 junction. This was explained on p.50. Current enters or leaves the plate in a way corresponding to changes in the input potential. This current is added to or subtracted from the steady current being supplied by the potential divider (R_1 and R_2). A fluctuating current reaches the base of the transistor causing the collector current (perhaps 200 times greater) to fluctuate proportionately. Although the variations in the minute current from a microphone would be too small to produce usable voltage changes across a resistor, the variation in the collector current produce appreciable voltage changes across R_3. For an increase of $1\mu A$ in the input base current produces an increase of 0.2mA in the collector current. This would cause an increase of 0.2V across R_3 if this were a $1k\Omega$ resistor. The effect is to decrease the output voltage by 0.2V. Note that although the transistor is a *current* amplifier we are using it to produce a *voltage* change. The amount of amplification depends upon the construction of the transistor. The gain of most transistors is between about 10 and 1000. Further gain can be obtained by connecting two transistors so that the collector current of one becomes the base current of the other. This gives the combination known as the *Darlington pair* Fig. 6.19. For small input

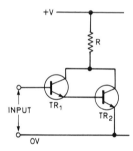

Fig. 6.19. Darlington pair

currents the gain of the pair is the *product* of their individual gains. If the gain of each transistor is 200, the gain of the pair is 40 000. Darlington transistors are produced that have the two transistors in one case with three terminals. Phototransistors commonly have a Darlington pair in the same case to amplify the tiny photocurrent, allowing the photo-Darlington to be connected into a circuit as a unit.

Two-stage amplifier

Another way of obtaining increased amplification is to connect two or more single transistor amplifiers one after the other, in *cascade*. Fig. 6.20 shows a straightforward amplifier circuit

suitable for use in an intercom system. The input current is
produced by a crystal (piezo-electric) microphone. Since this is a
high-impedance device, it may be connected directly to the
circuit across R_2. The potential divider R_1/R_2 provides the bias
current to make the transistor conducting. Variations in the
current from the microphone are amplified and produce variations in the collector current through R_3 and VR. The relatively
large changes of potential at point A are transmitted through C_2,
causing variation in base current to TR_2 and correspondingly

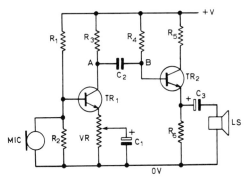

Fig. 6.20. Audio amplifier

greater variation in the collector current through R_6. As the
current in the resistor varies (in response to the original signals
from the microphone) the voltage at its 'upper' end varies
accordingly. The alternating voltage signal passes across C_3 and
actuates the loudspeaker to reproduce an audible signal.

It might be wondered why we do not simply join the collector
of TR_1 to the base of TR_2 using a piece of wire instead of a
capacitor. The reason is that the voltage levels at the two points
cannot be made equal. For satisfactory operation the voltage at
the collectors of TR_1 (point A) and TR_2 should lie about
half-way between the negative (0V) and positive rails, say at
4.5V, if +V is 9V. This allows the voltage to swing freely in
either direction. If the voltage were to be near to 0V or +V, then

it could not swing freely in one direction and waveforms would be clipped in that direction, introducing distortion. On the other hand, the voltage at the base of TR_2 (point B) must be low, not far above 0.6V, otherwise the transistor will be saturated. If we were to use a wire link, points A and B would necessarily be at the *same* voltage and the amplifier would not work. A capacitor can maintain a steady voltage difference between its plates. In this example it would be a little under 4V. This fixed voltage difference is a d.c. voltage. It does not prevent a.c. voltages being transferred across the capacitor. Thus C_2 functions as a *coupling* capacitor, allowing signals to be sent from one part of a circuit to another yet preventing the different operating voltages from interfering with each other.

The final point for consideration is the function of VR and C_1. When the potential at the base of TR_1 rises, base current is increased and so is collector current. The increased collector current through VR causes an increased p.d. between its ends. The potential at the emitter is *raised*. Raising the base potential automatically raises the emitter potential. Now, the factor that decides how much base current flows to TR_1 is the base-emitter p.d. If it happens that raising the base potential raises the emitter potential by the same amount, the base-emitter p.d. would be unchanged. The base current would be unchanged too. The signal would be damped out. This effect is known as *negative feedback*. The changes in output potential are fed back to neutralise changes in input potentials. If C_1 were not there, feedback could damp out the signal and no sound would be heard. The function of C_1 is to absorb *part* of the change, so that the amount of feedback can be controlled. By allowing a certain amount of feedback the gain of the amplifier is reduced, the advantage is that its gain is *constant*. It is not affected by the frequency and amplitude of the signal, or by temperature as it is when there is no feedback. The amount of feedback (and the gain) can be controlled by adjusting VR. Negative feedback improves the fidelity of amplification at the expenses of the *amount* of amplification. This is a reasonable exchange for it is easy, as is the case with this amplifier, to add a second or even a third transistor to regain the lost amplification.

FET amplifier

As explained on p.73, an FET has high input impedance. This makes it very suitable for use as the first stage of an amplifier that is to receive its input from a high impedance such as a capacitor microphone. Proper matching of impedance is essential for maximum power transfer between source and amplifier. In the circuit of Fig. 6.21, the current through R_2 causes the

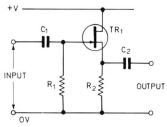

Fig. 6.21. Source-follower circuit using FET

source to be at a potential greater than 0V. The gate is held at 0V by R_1, so is negative to the source, as required for correct operation (p.72). The signal passes through C_1 to the gate, resulting in variation in the current through the transistor. TR_1 and TR_2 have low resistance so act as low-impedance sources to the next stage of the amplifier. Capacitor C_2 couples this stage with following stages, which may consist of junction transistors. The source follower does not produce any gain but allows connection between a high-impedance source and the low-impedance amplifying stages that follow. High-impedance capacitance microphones of the electret type frequently have an FET amplifier built into them, powered by a small dry cell. This matches the very high-impedance of these microphones to the medium-impedance amplifier or tape-recorder to which they are connected. When microphones are used at the ends of long screened cables, the high-impedance of the microphone acts in combination with capacitance of the screened cable to produce, in effect, a low-pass filter. The filter reduces the treble component of the signal, giving a muffled sound after amplification.

This type of distortion is reduced by use of the FET pre-amplifier in the microphone case, for its medium impedance output is not affected by cable capacitance to the same extent.

Operational amplifiers

Different types of operational amplifier i.c.s vary in their method of construction and in their characteristics but all have certain features in common. They all have two inputs, known as the inverting input ($-$) and non-inverting input ($+$), and have a single output. They operate relative to the 0V line, but they normally take their power supply from the positive line (say, $+15$V) and a negative line (say, -15V.). The connection to the power supply are omitted from Figs. 6.22-24. Also omitted are the connection to the offset null pins. These connections are made if it is important to provide fine adjustment of the output voltage. In many applications offset null adjustment is not required. It is omitted here in order that we may concentrate upon the amplifying functions of the i.c. itself. One important point about operational amplifiers (or *op amps,* as they are more commonly called) is that their inputs have high impedance. This is usually 2MΩ in bipolar op amps. Many op amps have a JFET or MOSFET input stage. For the latter the input impedance is 10^{12} ohms, or a million megohms. This is virtually *infinite input impedance*. The output impedance is very low, about 75Ω, so this is virtually *zero output impedance.*

As explained in the previous section, it is important to match impedance when power is to be transferred from one circuit to another. Often we need to transfer power from a high-impedance output to a low or medium impedance input. The op amp can readily act as an impedance buffer. In Fig. 6.22, the amplifier receives its input at the non-inverting terminal ($+$). This has high input impedance so is ideal for connection to, say, a capacitance microphone. Alternatively, if the intention is to measure voltage in a circuit where currents are small and resistances high, the op amp can be connected to any point in the circuit without the risk of it drawing so much current that the voltage level are altered. The output of the amplifier is fed back to the inverting input. The output of an op amp is proportional to the *difference* between the potentials at its inputs. In Fig. 6.22, if V_{IN} is 0V, V_{OUT} is 0V and this zero voltage is fed back to the inverting input. *Both* inputs are thus at 0V and the amplifier is in

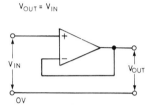

Fig. 6.22. Unity gain voltage follower

a stable state. If V_{IN} is raised to $+2V$, output begins to rise too. This puts up the voltage at the inverting input until it is equal to that at the non-inverting input; i.e. *both* are at 2V. The op amp then becomes stable with $V_{OUT} = V_{IN}$. It behaves as an amplifier with a gain of 1. The point of using the amplifier is for matching impedance. The output may be fed to low-impedance circuits or, for example, to a voltmeter with a low-resistance coil. The output can sink or source relatively large currents without serious effect on its output voltage.

Another key feature of op amps, not made use of in the previous example is that they have very high gain. Gains of over 100 dB (or 100 000 times, see p.127) are typical. This huge gain is known as the *open loop gain,* but is rarely made full use of. Generally the op amp is wired to feed back its output to the inverting input so that gain is markedly reduced. The reason is that the gain, known as *closed loop gain* is determined precisely by the values of the resistances in the circuit. Closed loop gain is not affected by frequency, temperature or by variation in the manufacture of the i.c. Fig. 6.23 shows how an op amp is connected as an inverting amplifier. When a positive voltage is applied to the inverting input the output swings negative. Since the non-inverting input is connected to the 0V line, and output is swinging negative we can think of R_1 and R_2 as forming a potential divider (p.114). Point A is positive, point B is negative, and some point along the chain must be at zero volts. Output

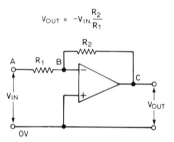

$$V_{OUT} = -V_{IN} \cdot \frac{R_2}{R_1}$$

Fig. 6.23. Inverting amplifier

swings negative until point B is at zero, for then both inputs are at 0V and the amplifier becomes stable. A current I flows along this potential divider and, since the inverting input has very high impedance, we can ignore any current flowing into it. So the current I flows from A to B and on to C with no change in value. Along R_1 the voltage drop is V_{IN} and

$$I = \frac{V_{IN}}{R_1}$$

Along R_2 the voltage drop is V_{OUT} and

$$I = - \frac{V_{OUT}}{R_2}$$

The negative sign is needed because V_{OUT} is negative.

Thus we can say

$$\frac{V_{IN}}{R_1} = - \frac{V_{OUT}}{R_2}$$

or

$$V_{OUT} = - V_{IN} \times \frac{R_2}{R_1}$$

The gain of this circuit is $-R_2/R_1$. For example, if we make $R_2 =$ 10MΩ and $R_1 = 10kΩ$, the gain is 10 000 000/10 000 = 1000. By choosing suitable resistor values we can obtain any required gain within limits.

In Fig. 6.23, the current flowing along R_1 is entirely diverted along R_2. Though the inverting input has high impedance it is always automatically brought to 0V by appropriate voltage changes at the output. It *behaves* as if it were an earth: we call it a *virtual earth*. Fig. 6.24 shows how we made use of this property in one of the circuits found in analogue computers. Numerical

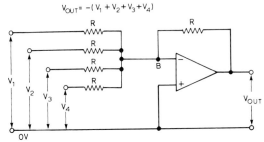

$$V_{OUT} = -(V_1 + V_2 + V_3 + V_4)$$

Fig. 6.24. Adder

quantities are represented by different voltages. We may have four quantities, represented by four voltages V_1, V_2, V_3 and V_4. If these are applied to the four inputs of the adder circuit, currents proportional to these voltages flow along the *equal* input resistors. The combined currents flow along the feedback resistor (also equal to R). To bring the potential at B to zero, output must fall by an amount equal to the *sum* of the four input voltages. Circuits such as these can be used for adding any reasonable number of quantities together (p.157). There is not enough space here to describe any more of the applications of op amps. As well as their obvious uses as straightforward amplifiers, buffers and level detectors, these versatile devices have uses as subtractors, integrators, differentiators, current-to-voltage convertors, oscillators and precision rectifiers. They illustrate so well the way in which integrated circuits have become the building blocks of electronics.

7 Test Instruments

The moving-coil meter

If a coil carrying a current is placed in a magnetic field a force will act upon the coil. An instrument using this principle is shown in Fig. 7.1. The coil L is wound on a former of aluminium over a cylindrical iron core. The former is mounted on pivots so that it

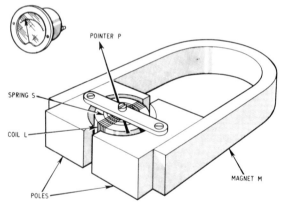

Fig. 7.1. The moving-coil meter

can rotate between the poles of the permanent magnet M. There is only a narrow air gap between the poles and the coil former, so that the magnetic field is radial and uniform for any position of the coil. The pointer, P, carried by the coil, is retained by the phosphor bronze spring S. When a current passes through the coil it tends to turn on its axis against the opposing force of the spring until it takes up an equilibrium position which is proportional to the current flowing through the coil.

Measurement of current, voltage and resistance

The basic moving-coil ammeter arrangement that we have just
described forms the basis of a number of other meters. By
suitable modification it can be used to measure voltages and
resistances. And its range as a current measuring meter can be
greatly extended so that it will measure a number of different
current ranges.

If a special low-resistance shunt is placed across the meter–see
Fig. 7.2 (a)–a certain current–depending on the resistance value
of the shunt–will flow through the shunt. The meter will then
take only a proportion of the total current flowing through the
whole arrangement. Thus a meter movement which is fully
deflected when 1 mA flows through it can be made to deflect
fully when 10 mA flows through a circuit, 9 mA going through
the shunt. By arranging for shunts of various values to be placed
across the meter, the meter will indicate currents of widely
different magnitudes. Selection between different current ranges
is easily effected by switching. The dial of the meter will usually
have the same number of scales calibrated on it as there are
ranges in the meter.

So far we have been considering the measurement of direct
current. To measure alternating current a rectifying diode is
connected in series with the meter to turn the a.c. into d.c.

To measure voltage, the meter is connected to the circuit
under test via a series resistor–see Fig. 7.2 (b)–of appropriate
value. A small current will flow through this series resistor, and
as, in accordance with Ohm's law, this current is proportional to
the voltage across the resistance, the scale of the meter can be
calibrated in volts.

Including a battery in the meter enables it to measure resistance. Fig. 7.2 (c) shows the arrangement. The battery current
flows through the resistor under test and the meter: once more,
following Ohm's law, the meter can be calibrated in ohms.

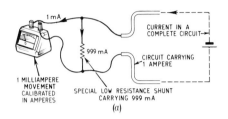

1 mA

999 mA

CURRENT IN A
COMPLETE CIRCUIT

CIRCUIT CARRYING
1 AMPERE

1 MILLIAMPERE
MOVEMENT
CALIBRATED
IN AMPERES

SPECIAL LOW RESISTANCE SHUNT
CARRYING 999 mA

(a)

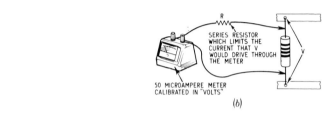

R

SERIES RESISTOR
WHICH LIMITS THE
CURRENT THAT V
WOULD DRIVE THROUGH
THE METER

V

50 MICROAMPERE METER
CALIBRATED IN "VOLTS"

(b)

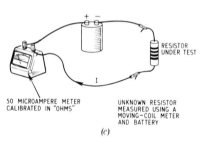

+ −

RESISTOR
UNDER TEST

I

50 MICROAMPERE METER
CALIBRATED IN "OHMS"

UNKNOWN RESISTOR
MEASURED USING A
MOVING-COIL METER
AND BATTERY

(c)

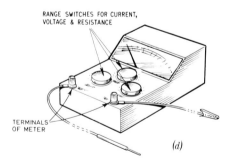

RANGE SWITCHES FOR CURRENT,
VOLTAGE & RESISTANCE

TERMINALS
OF METER

(d)

Fig. 7.2. Tests using a multimeter

The FET voltmeter

The multimeter can be used for most measurements but under certain conditions gives misleading results. For example, it may be suspected that the voltage at the base of a transistor is wrong. On measurement with the voltmeter section of the multimeter a very low reading is obtained. The person testing the circuit may immediately suspect a fault whereas in fact none exists. What has happened is that the voltage is reduced to a near zero level by the relatively heavy current drain of the meter. The resistance of the voltmeter is in fact lower than the source resistance of the circuit under test and thus the meter has short-circuited the biassing resistor.

In Fig. 7.3 the current flowing through R_1 would normally flow on through R_2, except for a small fraction flowing to the base of the transistor. With $R_1 = R_2$, the potential there would be close to

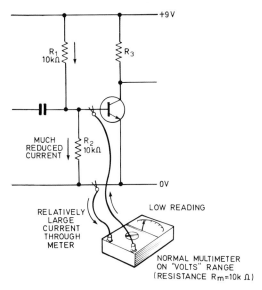

Fig. 7.3. The multimeter can give misleading results in some instances

4.5V. With the voltmeter attached to the circuit, as shown in Fig. 7.3, the resistance of the meter (10kΩ) is in parallel with R_2. Half of the current from R_1 is diverted through the meter. The combined resistance of the meter and R_2 is only 5kΩ. This means that the voltage at the base of the transistor falls to 3V. The meter reads 3V, yet the voltage at that part of the ciruit is really 4.5V *when the meter is not connected to the circuit*. The effect is even more pronounced if R_1 and R_2 have greater value: the meter may take almost all the

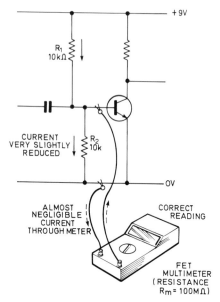

Fig. 7.4. The FET multimeter used for accurate voltage measurement

current, bringing the apparent base voltage down almost to zero. The FET meter has a very high input impedance, typically up to 100MΩ or even more. The current flowing through the meter is negligible. In parallel with R_2, the combined resistance is 9999Ω. This differs so little from the resistance of R_2 alone that the base potential is virtually unaffected, and the voltmeter shows the true value.

There are many other circuits in electronics that reveal this basic deficiency of the simple multimeter. Because of this a more elaborate meter, the FET voltmeter, is needed. The meter incorporates a battery-powered amplifier with an FET input stage. This may be a JFET or MOSFET and the amplifier may possibly be an op amp i.c. The output of the amplifier is fed to a meter of the normal kind or to a digital display. The essential point is that the FET has such high input impedance that it draws negligible current from the circuit under test.

In *digital* multimeters the output of the amplifier is not fed to a moving-coil meter but to electronic circuits that indicate the voltage, current or resistance on a series of 7-segment displays. These may be either the LED type or an LCD (p.91). Such meters often include extra facilities such as automatic selection of range and the ability to operate independently of the polarity of the input. Provided that they are well designed and properly calibrated, such meters measure with a very high degree of precision. In spite of this, some engineers prefer the moving-coil meter for measuring fluctuating voltages and currents. Useful information can be learned by watching the way the needle moves across the scale. It is far from easy to extract this information from a set of rapidly changing digits.

The oscilloscope

One of the most useful 'tools' in electronics is the oscilloscope, the heart of which is a cathode-ray tube. A typical tube is shown in Fig. 7.5. To provide the necessary voltage to operate the tube, a long resistance chain may be used. This chain, which may have a total resistance of several megohms, has a voltage of two or three thousand volts applied across it called the e.h.t. supply (extra high tension).

The cathode-ray tube consists of a specially shaped glass tube evacuated of all gas, and containing a number of electrodes. The first of these is the filament which heats the cylindrical cathode to the correct working temperature. The electrons emitted by the cathode are accelerated by the main anode and, as the anodes and the grid are in the form of cylinders, they pass through and eventually hit the screen at the end of the tube. The screen is covered by a phosphorescent material which glows

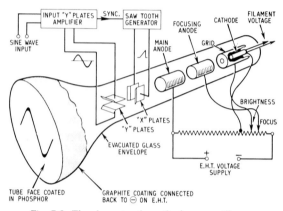

Fig. 7.5. The elements of a cathode-ray oscilloscope

when hit by the electrons, the number of electrons that hit the screen determining the brightness of the glow. The grid is connected to an appropriate place on the resistance chain and its potential relative to the cathode can be altered by varying this position. As the voltage on the grid determines the number of electrons that pass through it, this voltage determines the brightness of the image on the screen. It is usually negative relative to the cathode.

After leaving the grid cylinder the electrons pass through a focusing cylinder. This is made negative with respect to the grid and tends to 'squeeze' the electrons into a pencil-thin beam, although the grid cylinder also helps in this process. Thus the result of these two electrodes is a sharp, finely focused beam of electrons. The focusing cylinder is also connected to the resistance chain and its position is made variable to give some control of focus. The electrons are now travelling towards the second cylinder, or main anode, which is highly positive with respect to the cathode and has attracted the electrons away from the cathode in exactly the same manner as the anode of a conventional valve. The grid cylinder acts very much the same way as the grid of a valve except that in the cathode-ray tube it is in the form of a cylinder and not a mesh.

After passing through the main, or final anode, the electrons are travelling at a very high velocity. After impinging upon the face of the cathode-ray tube the electrons eventually find their way back to the cathode via a graphite coating on the glass tube. On striking the end of the tube this current gives up enough energy to the phosphor coating to make it glow visibly: if the electron beam is stationary and not being deflected the tube will have a small bright spot stationary in the centre of the tube face. The colour of the spot may be green, blue, orange and so on, and it may have short persistence, depending upon the phosphor used by the manufcturer and the final purpose of the tube. The term phosphor has nothing to do with the chemical phosphorus –it is used in this connection to describe any material used to coat the face of the cathode-ray tube because it glows when activated by electron bombardment. The term persistence is self-explanatory. A long-persistence tube may glow for several seconds or more after bombardment. In measurement work in electronics short-persistence tubes are used to prevent confusion of the beam movements, which may be changing over a period of milliseconds.

'X' and 'Y' plates

To produce a high-velocity, properly focused electron beam is the function of the electrodes along the axis of the tube, but before the beam reaches the face it has to pass through two sets of deflector plates, the 'X' plates and the 'Y' plates. If a sawtooth voltage is applied to the 'X' plates the beam will be drawn across the face of the tube and a visible line will be produced on the tube face. With the correct persistence value and the correct repetition frequency of sawtooth voltage the line appears to be permanent. When the sawtooth pulse reaches its maximum value it rapidly returns to zero and the spot will fly back to the beginning of its trace–so rapidly as to produce no glow at all. Therefore the important part of the sawtooth waveform is the relatively slow slope. The circuit producing the 'X' plate deflection voltage is called a timebase circuit.

The voltage to be observed and measured is applied to the 'Y' plates. If this voltage is a sinewave, i.e. alternating, it will appear on the face of the cathode-ray tube as shown in Fig. 7.5, the trace due to the sawtooth voltage on the 'X' plates being varied in a sinusoidal fashion by the voltage on the 'Y' plates.

Oscilloscope input

The input to the oscilloscope may be very small and in this case amplification is necessary before it can be used to produce a noticeable deflection on the tube face. The amplifier which does this is called the 'Y' plate amplifier, and a little of its output is fed to the 'X' plate sawtooth circuit for synchronizing the trace so that it begins at the appropriate time to allow the voltage on the 'Y' plates to appear correctly related to the 'X' plate scan.

Signal and pulse generators

A signal generator is a low-impedance source of a range of frequencies, its output being a fixed, or variable and calibrated, sinusoidal one. If the output is calibrated the generator is usually more expensive and is used mainly in laboratory work. The simpler and cheaper fixed-output instrument is used mainly for qualitative checks on circuits.

The term low impedance in this connection means that if the instrument is regarded as a generator of energy, which indeed it is, its internal impedance is low so that when current is drawn from it by the external load (the circuit under test) very little voltage is lost internally across the generator impedance. The calibrated-output generator must of course have a very low internal impedance in order that its output, once set, will not vary as the load current fluctuates during adjustments.

The cheaper type of signal generator is generally used, rather like a bell-buzzer, for continuity tests on circuits, to see whether signals are passing through equipment rather than to enable precise measurements of signals at particular points in the circuits to be made.

Basically the signal generator consists of an oscillator which can be tuned over a series of appropriate ranges. The output at the chosen range is then fed via some circuit such as an emitter-follower to give a low-impedance output. The range of frequencies covered may be from a few kilohertz to a few hundred megahertz. To test television equipment a video oscillator is used. This is a signal generator that gives a square wave output as well as sinusoidal output at frequencies ranging from a few cycles per second up to several megahertz per second. The output may be calibrated or not as in the case of the r.f. signal generator.

In addition to generators for video work there are many types of pulse generator available which offer various shaped pulses of many different lengths and having many different mark-space ratio settings. A typical output might be a square pulse with a rise time of 0.25 μsecs and a length of 5 μsecs occuring at space intervals of 20 μsecs–a pulse repetition rate of 40 pulses per second.

8 Computer Electronics

Logic circuits

In this age of computerization and automation, logic circuits are playing an ever-increasing part in our lives. In almost every home there is at least one device that operates by digital electronic logic. There are many forms of logic, but the one most suitable for translation into electronic form is *binary logic*. In binary logic we constantly deal with *two* states. A statement is true or untrue: there are no half-truths. A digit is 1 or 0: there are no fractions or other values. A transistor is either fully on or fully off: it switches rapidly from one state to the other. A voltage is either high or low: intermediate values have no meaning.

The on-off, high-low characteristics of binary logic mean that we can more easily design electronic circuits that can model logical statements and numerical values with absolute accuracy. Arithmetical operations are performed using binary numbers as explained on p.158. Logic is performed by a number of standard operations including these:

NOT If A is true, B is NOT true; if A is NOT true, B is true. Example: If the door is open it is NOT closed: if it is closed it is NOT open.

AND If A is true AND B is true, then C is true; otherwise C is false. Example: If the door is open AND I am in the room then I am able to leave the room by way of the door; otherwise I am not able to leave.

OR If A is true OR B is true, then C is true; otherwise C is false. Example: If the door is open OR the window is open, I can enter or leave the room, otherwise I cannot.

The operations can be represented by truth tables in which '1' represents 'true' and 'O' represents 'false':

NOT		**AND**			**OR**		
INPUT	OUTPUT	INPUT		OUTPUT	INPUT		OUTPUT
A	B	A	B	C	A	B	C
1	0	0	0	0	0	0	0
0	1	0	1	0	1	0	1
		1	0	0	1	0	1
		1	1	1	1	1	1

Operations such as AND and OR can be extended to include more than two conditions: eg. If A is true AND B is true AND C is true AND D is true, then E is true; otherwise E is false.

It is a simple matter to perform these operations using ordinary manually-operated switches (closed=true, open=false) with lamps to indicate the logical outcome. For calculators, clocks and computers we need full electronic switching at high speed, combined with reliable action, freedom from interference by occasional voltage fluctuations and 'spikes,' and outputs that are firmly high or low. Logic circuits that meet these specifications are made in integrated form and we will now consider some of them.

Transistor-transistor logic

This type of circuit, usually known as *TTL,* superseded an earlier form, diode transistor logic (DTL). TTL circuits all operate on a +5V d.c. supply. The circuit element that performs a single logical operation is known as a *gate.* Fig. 8.1 shows the circuit of a commonly-used gate that performs the operation NAND. This is short for NOT AND and, as can be seen from its truth table, this gate gives the inverted output of an AND gate. The circuit includes an unfamiliar component–a transistor with two emitters. The transistor can conduct when *either* one *or* both emitters are connected to ground. If *both* are connected to +5V (A *AND* B) the transistor does not conduct, making output at C go low

(0V). With all other combinations of input, C is high (+5V). Such a single circuit does not occupy much space on a silicon chip so four independent gates are fabricated in one d.i.l. package, the whole lot costing little more than a single transistor.

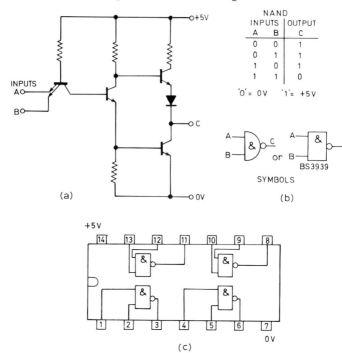

Fig. 8.1. TTL. (a) circuit of a two-input NAND gate. (b) Its truth table. (c) Symbols for NAND gates. (d) The 7400 i.c., which contains four two-input NAND gates, with shared power supply

TTL comprises a large family of several hundred different i.c.s. Many of these consist of relatively simple gates, but others employ MSI and LSI (p.111) to build complex counters and registers on a single chip. The chief advantage of TTL is its high speed, making it suitable for use in computers. It needs a regulated supply voltage, which should be between 4.75 V and

5.25 V and should not carry electrical noise that will interfere with the operation of the circuit. One disadvantage is that TTL requires fairly large currents. Even a relatively small system consisting of half-a-dozen i.c.s can require a supply of 1A, which means that the use of TTL in battery-powered equipment is severely limited.

Another, slightly more expensive version of TTL is the low-power Schottky TTL. This requires only one fifth of the power of standard TTL, yet operates at twice the speed. Most standard TTL devices are also available in Schottky form.

CMOS logic

Although CMOS logic operates more slowly than TTL, this is no disadvantage for very many applications. Its chief advantages are that it has very low-power consumption (making it ideal for battery-operated portable equipment), it can operate on any voltage in the range 3V to 15V, and it is relatively unaffected by electrical noise on its supply lines. A regulated power supply is not required. Another important feature of CMOS is that the individual components can be made so small that thousands of components can be accommodated on a single small chip. This makes CMOS the obvious choice for large-scale integration and

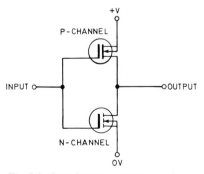

Fig. 8.2. Complementary MOS gate that performs logical NOT

very large scale integration. Consequently the CMOS range of i.c.s includes a 14-stage shift register and other similarly complex devices that are not available in the TTL range.

CMOS logic is based on the complementary pair of enhancement mode transistors shown in Fig. 8.2. One is a p-channel transistor (see p.75) and the other is an n-channel transistor. When the input is grounded (0V), the p-channel transistor turns off, so acts as infinite resistance. The result is that output swings high. Conversely, when input is high, the p-channel transistor is off, the n-channel transistor is on and output swings low. This complementary pair acts as a NOT gate. Note that since inputs to CMOS transistors have exceedingly high impedance, virtually no current is required to drive the gate. Secondly, no current flows except while the gate is changing state. This means that operating current is exceedingly low, especially at low frequencies or in a quiescent circuit. Fig. 8.3 shows a NAND gate, the

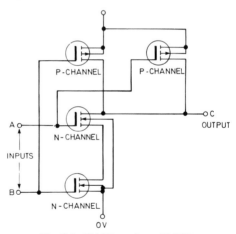

Fig. 8.3. CMOS two-input NAND gate

CMOS equivalent of the TTL gate of Fig. 8.1. If both inputs A and B are high, both n-channel transistors are on and output is connected to 0V (low) through the two transistors in series. The p-channel transistors are off, isolating output from +V. If either one or both inputs are low, either one or the other or both n-channel transistors are off, disconnecting output from 0V. At the same time one or more of the p-channel transistors is on, so output goes high. Note that this gate requires only four components, the transistors, against nine components in the corresponding TTL gate.

Another useful CMOS gate is the transmission gate or bilateral switch (Fig. 8.4). When the control input to this gate is made low, the n-channel transistor is off. The p-channel gate is off too,

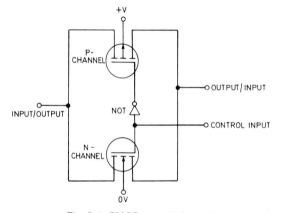

Fig. 8.4. CMOS transmission gate

for this is receiving a high input through the NOT gate. Thus input and output are isolated from each other. Only a leakage current of a few nanoamperes can flow. When the control input is made high, both transistors come on. Current can now flow in either direction through the transistors and the resistance between input and output terminals is only about 80Ω. The gate acts as a resistor of low value. Analogue signals can be transmitted in either direction. The transmission gate allows analogue signals (eg. audio signals) to be switched under logical control.

CMOS and TTL are the most widely used families of logic circuits. Apart from these there are several other families with special applications. These are known by abbreviations such as

I²L (Isoplanar Injection Logic), HLL (High Level Logic), HNIL (High Noise Immunity Logic) and ECL (Emitter Coupled Logic). HNIL, and several of the other systems are slow in operation, which makes them less affected by transient pulses and other types of noise in the system. They are specially suited to industrial applications where an electrically noisy environment would make it impracticable to use TTL. Their slow speed is of no disadvantage in most of these applications. ECL, on the other hand, is designed for extra-high-speed operations. A new type of high-speed logic is described in the next section.

Josephson logic

The Josephson switch can be used as the basis of all types of logic gate. It does not depend on semiconduction, even though silicon is used as the substrate, but on *superconduction.* Superconduction occurs in metals when they are cooled below a certain temperature close to absolute zero. Then the metal has zero resistance. Once current has started to flow it can continue to flow indefinitely. Flip-flops and other memory devices can store information in the form of currents that will circulate for ever unless they are cancelled. One of the advantages of superconduction is that because there is little or no resistance there is little dissipation of power. A whole computer would require only 7 watts, campared with the thousands of watts required by a normal computer. The Josephson junction (Fig. 8.5), consists of two superconductors separated by a very thin layer of insulating material. The insulating layer is of lead and indium oxides and is about 5 nanometres thick. When electrons meet this layer they are able to tunnel through the layer, as in a tunnel diode (p.66).

If an increasing current is applied to a Josephson junction, there is at first no increase in voltage across the junction, for the whole junction is a superconductor, *including the insulator.* When a certain current is reached (I_{MAX}), superconduction ceases and a voltage develops across the junction. If the current is now reduced, superconduction is not resumed until current falls below a certain value (I_{MIN}). Thus a short high-*current* pulse can turn it on, and a short low-*current* pulse can turn it off. Here

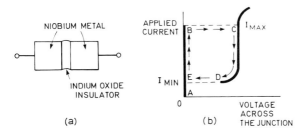

Fig. 8.5. The Josephson junction. (a) Structure. (b) Changes in voltage across junction as current is increased from zero (A) to I max (B- then jump to C) and back to I min (to D then jump to E)

we have a *current*-controlled switch. The current levels at which switching occurs depend on the magnetic field in the region of the junction. Suitable magnetic fields can be produced by passing current through a conductor close to the junction. By varying *this* current we can alter the magnetic field and lower the switch-on level or raise the switch-off level. In this way the junction can be controlled by a current in another circuit, just as a transistor switch is controlled by varying its base current.

A computer built from these junctions will have high capacity owing to the extremely rapid rate at which the junction can change state. The computer will operate at about 20 times the rate of a normal computer. Yet the computer *must* be small in size for, if cycle time is only 1 or 2 nanoseconds, the electrical signals can travel only 15 or 30 cm in that time. It would be difficult to synchronise the action of a computer of large size. The Josephson computer will consist of about 10 000 chips compactly mounted in a 14 cm cube and immersed in liquid helium. Despite its small size, and low operating current its performance will be superior to any conventional computer.

The logic elements of the Josephson computer have been built and tested but the complete computer is still in development. Although it depends on an entirely different technology, the fundamental computing principles on which it works are the same as those of the commonly used computers to be described in the next section.

Computers

Although this chapter is headed 'Computer electronics' we shall not go into more than the briefest details of the construction and operation of computers here. This subject is more fully dealt with in the companion volumes, the *Beginner's Guide to Computers* by T. F. Fry and the *Beginner's Guide to Microprocessors* by Andrew Parr. Here we are more concerned with the way electronics is used on computing. Whether we are dealing with the large mainframe computer, the smaller but very powerful minicomputer, or the popular and very capable microcomputer, the general organization and the electronic principles of a digital computer are the same.

Before going on to consider digital computers, it is worth looking briefly at another type of computer that is entirely different, the *analogue computer*. This is designed not for logical operations but for performing complex mathematical calculations. The values that enter into the calculations and the results obtained are represented on a continuous scale by a number of voltages. Each voltage is the *analogue* of the quantity it represents. The computer consists of an array of units each based on operational amplifiers. There are adders, multipliers, integrators and other units that can be connected together in a variety of ways. They are connected so as to model the physical event that is to be studied. A trivial example is shown in Fig. 8.6. Water flows into a tank at rate *a* litres per second, and flows out at *b* litres per second. How does the height of water, *h*, vary with time? We can see that: rate of change of *h* = constant × (rate of inflow − rate of outflow).

This is represented mathematically in equation 1. To find the height after any given time we integrate this equation, adding h_0, the height of water in the tank at the beginning, to get equation 2. But although *a* is fixed, *b* gets greater as height increases. This gives us equation 3. Equations 2 and 3 are modelled by various circuits based on op amps, as in Fig. 8.6. The inputs to the circuit are the values of voltages *a, k,* and *k,* which can be set by high-precision potential dividers. The value of *h* is calculated and fed back so that *b* can be calculated. An initial voltage h_o is applied to multiplier 1 to take into account the water in the tank at the beginning, When the circuit is switched on, the output h rapidly settles at value h_o, as the various parts come into equilibrium. Then h rises (assuming a is greater than b) at a rate corresponding to the rate at which water would rise in a *real* tank

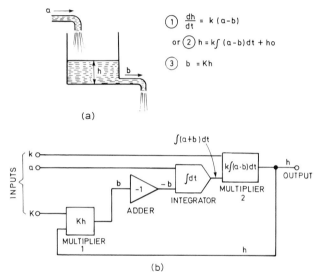

(1) $\frac{dh}{dt} = k(a-b)$

or (2) $h = k\int (a-b)dt + ho$

(3) $b = Kh$

(a)

(b)

Fig. 8.6. Analogue computer. (a) Problem and equations. (b) Electronic solution. The adder is an op-amp connected as an inverting amplifier

with *real* water flowing in out of it. The rate at which this happens in the computer is set by resistance-capacitor timing circuits in the integrator. We can make the computer run faster or slower than real events, or at the same rate, as we wish. Output h may be fed to a voltmeter or to a graph plotter. We can vary the factors a, k and K at will to see what happens under different sets of conditions.

An analogue computer would not normally be used to solve such a simple problem as that described above. It could be used, for example, to predict the flow and levels of water in a complex irrigation system. Analogue computers are good at solving complex equations. Often they are built with one particular type of problem in mind. For example they can be used to control production processes where complex calculation is involved. In this situation they may well be cheaper and more suitable than a digital computer.

Binary digits

As explained on p.148, data and instructions are presented to computers in binary form:

Decimal number	Binary number
0	0000
1	0001
2	0010
3	0011
4	0100
5	0101
6	0110
	… and so on

In the table above, each number is represented by four binary digits (called *bits,* for short). Computers usually operate with 8-bit numbers, though some of the larger ones use 16-bit numbers. An 8-bit number is usually refered to as a *byte.* This unit is used for describing the size of the memory of a computer, though we normally speak of kilobytes or megabytes, or even gigabytes when referring to the memory store of a mainframe computer. Since bytes too are counted in the binary system, a kilobyte is not 1000 bytes, as might be expected, but 1024 bytes. This figure is 2^{10} (in decimal, or 10000000000 in binary) the nearest 'round number' in the binary scales.

To a computer, a byte may represent numerical data (an actual binary value), alphabetic data (a letter of the alphabet or a symbol in binary code), an address of a section of memory where data is to be stored on from which data is to be read, or an instruction to perform a particular operation. Exactly what any given byte means depends on its position in the program. The

central processing unit (see Fig. 8.7 and next section) knows
what to expect and interprets each byte accordingly.

The computer recognises only high or low voltages, which we
represent by '1' or '0.' A byte can be written out as an 8-digit
number, for example 01001011, but this is a cumbersome system
to use. It is easy to make mistakes, for 01001011 looks very like
01001101 yet might mean something entirely different to the
computer. Programmers use *hexadecimal code* for writing out
programs on paper. This uses 16 characters, the numbers 1 to 9
and the letters A to F. 0 to 6 are as tabled above; the remainder
are:

Hexadecimal number	Binary number
7	0111
8	1000
9	1001
A	1010
B	1011
C	1100
D	1101
E	1110
F	1111

To represent the binary number 01001011 we split it into two
halves 0100 (=4) and 1011 (=B). It is written as '4B' in
hexadecimal. Similarly, 01001101 is written as '4D.' The hexadecimal system is used in writing programs in machine code (p.171)
and when entering such programs at a computer keyboard.

The parts of a digital computer

The essential parts of a digital computing system are shown in
Fig. 8.7. A mainframe computer has a larger memory store than
other types of system and usually has a more varied assortment
of input and output devices. A microcomputer has most of the
system (except for memory and the input and output devices) on
a single chip. Apart from such differences of scale, Fig. 8.7 can
be applied to all kinds of computer. Briefly, the functions of the
parts are as follows:

Clock: generates timing pulses at the rate of several megahertz to synchronize the activities of all parts of the system.

CPU: this is where calculations and logical operations are performed. It consists of several sets of *registers* (usually sets of eight or sixteen circuits of the flipflop type (p.120) for holding data, instructions or addresses in binary form. One of these, the *accumulator,* holds data on which operations are currently being performed.

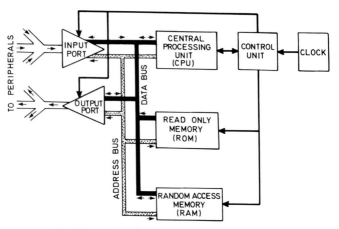

Fig. 8.7. The essential features of a computing system

Input port: receives data, addresses or commands from peripheral devices such as a keyboard, teletype, punched tape reader, punched card reader, or magnetic data store (see later). In industrial or scientific applications, input may come directly from electronic sensors or measuring circuits. Input may arrive from the output port of other computers either directly or by way of the public telephone system. In the latter case a device known as a *modem* is used to link the computers to the telephone lines. A *light pen* provides input that allows the user to 'draw' on the screen of the VDU (see below). The user may also write or draw on a graphics table; input from this is converted by the computer to diagrams and words that are displayed on the VDU.

Output port: often the same i.c. is used as for input, it being possible to switch gates to function as input or output as required. The output peripherals may include a printer or teletype, a visual display unit (VDU, similar to a television screen), a modem, a magnetic store, or a plotter (plots graphs, etc). Industrial computers may be connected directly to the machinery under their control.

Buses: these consist of sets of connecting lines running in parallel. An address bus may consist of up to 16 lines, each carrying one bit of a 16-bit address. The data bus usually has 8 lines (one byte), since this is the amount of data normally transferred at one operation. As the diagram shows, data buses are bidirectional. It would make wiring far too complicated if every section of the computer had its own set of eight wires joining it to every other section. The buses are communal: each part of the computer can take its turn (under the direction of the control unit) to send or receive messages along the address or data buses. When a part is *not* delivering a message to a bus it must not be allowed to interfere with the message being sent by other parts. We cannot arrange to have switches to disconnect its outputs from the bus. Instead we use a gate with *tri-state output*. Fig. 8.8 shows the output stage of a CMOS tri-state gate. If E is high (and Ē low) the output is 'enabled' because *both* transistors of the complementary pair are conducting. The output of the

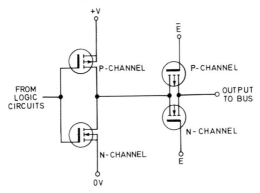

Fig. 8.8. CMOS tri-state output

gate is 'high' or 'low' depending upon the logic before it. The gate can sink or source current, for it is connected either to the positive line or the 0V line. It behaves just as an ordinary gate and can send signals to the bus or receive signals from it. When the device is disabled, by making E low (and $\bar{E}$ high) both transistors are off. This disconnects output completely: it now has very high impedance. It can neither send nor receive signals. It has no effect on signals passing along the bus between other pairs of devices. The enabling or disabling of outputs is under the control of the control unit; this is how it controls the data flow with the computer and between the computer and its peripherals.

Read only memory; this consists of one or more i.c.s containing numerous (several thousand) memory units of which each has an address–usually an 8-digit binary number. When the address of a given memory unit is fed to the address inputs of the ROM, the outputs of the ROM go high or low to indicate another 8-digit number. This is the information stored at the address. When first manufactured the ROM produces all-high ('1') outputs from all addresses. Before being placed on the computer it is programmed, usually by 'blowing' fusible links by passing a high current through them. Each blown gate thereafter produces a low ('0') output. In this way the ROM can be made to hold the sequence of operating instructions for the CPU. It is the computer's 'working handbook.' In operation, the CPU addresses each memory unit in turn (*via* the address bus), reads the instructions it contains (*via* the data bus), and acts as instructed. It is possible to substitute different ROMs to improve the CPU's performance or to instruct it how to perform specialised tasks. The computer cannot alter the contents of the ROM itself, but there are ROMs that can be altered or reprogrammed outside the computer. These usually operate by the storage of electric charge on the gates of certain transistors. MOSFETs (p. 75) have an insulated gate that can retain a charge for several months or years if disconnected. Programming consists in charging the appropriate gates. Later the ROM is reprogrammed by electrical means (electrically alterable ROMs, or EAROMs) or by exposing the i.c. to ultra-violet light. The latter type of i.c. has a quartz

window to allow ultra-violet light to enter. Its action is to erase
all data stored, hence the name EPROM (erasable programmable ROM). After this they can be reprogrammed electronically.
Random access memory: The CPU cannot only read data from
this kind of memory but can write data into it as well. Essentially
it consists of numerous flip-flops (p.120) that can be set (giving
output '1') or reset (giving '0'). Usually these flip-flops are
arranged in banks of eight, to store or provide 8-digit binary
numbers. The RAM may be used to store programmes that the
CPU is to execute, to store data upon which it is to work, or to
act as a 'scratch-pad' in which it can temporarily store the results
of various stages in its calculations. The essential point about
RAM is that the CPU can store or read instructions or data at
very high speed. This is essential if it is to be able to make full
use of its abilities to calculate and perform logic at high speed.
However an array of RAM i.c.s cannot store the masses of data
required by a computer (for example the names and personal
details of employees of a large firm) and a magnetic back-up
storage is also required. In addition, data in RAM is lost when
the power is switched off. Any data that is to be retained must be
transferred to magnetic storage before the computer is switched
off.

Large capacity magnetic stores

In many computer applications there is a need for bulk storage
systems that can store many millions of pieces of digital information. Although many different ways of doing this have been
devised, the method that is almost universally used today is to
record the information on thin magnetic tape or film.

The information is stored by changing the direction of magnetization of small areas of the magnetic film as shown in Fig. 8.9.
In this figure, a binary 1 is represented by magnetizing a small
area in the one direction, and a binary 0 by magnetization in the
opposite direction.

Information is written on the magnetic surface by a write head
which works in much the same way as the recording head on an
ordinary domestic tape recorder. A current is passed through the

coil setting up a strong magnetic field in the narrow air gap. If the magnetic film passes very close to the air gap, the field magnetizes the film in one of the two possible states, depending on the direction of the current in the write coil (Fig. 8.9). The use of only two magnetization states (corresponding to the

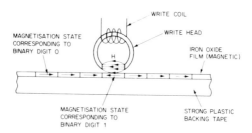

Fig. 8.9. Writing on to magnetic film

binary digits 0 and 1) differs from a tape recorder in which a continuous spectrum of degrees of magnetization directions is used (analogue recording).

Naturally, it is no use having a storage medium containing information which cannot be read out, and so the next problem is to read the stored digits. Here the system employed is very similar to the playback head of a tape recorder except that only two magnetization states have to be detected. The read head is just like the write head and in some systems the same head is used for both writing and reading. The magnetic medium is moved at a uniform speed past the read head close to the air gap. Any change in film magnetization alters the flux in the gap and induces a current pulse in the read coil. This pulse can be amplified and displayed or used to perform some other task in the computer.

To recover the recorded information it is necessary to include timing information. This is most easily done by recording a pulse in each digit position on a timing track parallel to the data information tracks.

There are three important systems which use this method of storing and retrieving information. They are magnetic tape, magnetic drum, and magnetic disc. All these systems store information efficiently and accurately, and the choice between them depends on the task performed by the computer, the cost of the systems, and the time that can be allowed for the retrieval of information from the store. For example, when performing complex mathematical computation it is necessary to have data available very quickly indeed, whereas in an airline booking system a delay of a minute might be acceptable.

Magnetic tape

The magnetic tape system is similar to the domestic tape recorder from which it is developed. It is, however, much more sophisticated and costly. The tape transport mechanism is designed to wind the tape at high speed when searching for information, and stop it almost instantly (without breaking the tape) when the required data has been located. The transport mechanism must also run the tape through at a constant speed for reading and writing data.

Up to 30 tracks can be recorded in parallel on a standard 25.4 mm (1 in.) tape. Most of these tracks store data, but some hold timing information and also information to identify particular blocks of data so that they can be found quickly when required. A library of such tapes may contain an almost unlimited amount of data.

One of the drawbacks of magnetic tape is that the time taken between the system receiving a request for data and the start of reading the data (known as the access time) may be several minutes if the information is at the far end of the tape, because the whole tape will have to be rewound. The average access time for tape units is about one minute.

The great popularity of cassette tape recorders has lead to their use with microcomputers. Data is recorded on tape and read using an ordinary domestic cassette recorder. Access time is slow and there is no provision for rapid winding of tape under the control of the computer. This is the most frequently used method of data and program storage for the home enthusiast and small business user. A wide range of pre-programmed tapes is available covering business programs, games and educational programs.

Magnetic drums

The long access time of magnetic tape systems can be considerably reduced by using a magnetic drum. In this system a magnetic film is wound round the perimeter of the drum, Fig. 8.10 (a). Recording tracks are placed parallel across the drum with a read/write head for each track. As with magnetic tape, some tracks are devoted to storing data, and others to timing and address information to aid data retrieval.

Because the drum can be rotated at high speeds (up to 15 000 rev/min) and as the required data block is, on average, only half a revolution away from the head, the access time is only a few milliseconds. The main disadvantage of a drum store is that it can only hold about ten million binary digits which is less than a single magnetic tape. Also they are bulky and cannot be easily changed.

Magnetic discs

In this system, information is stored on discs which look like gramophone records. However, instead of the information being stored by cutting a groove in the material, the disc is coated on both sides with a thin layer of iron oxide and the data is stored magnetically on concentric tracks in the same way as on magnetic tapes and drums Fig. 8.10 (b).

The combined read/write for each disc side can be moved radially to select the required track, and the disc is rotated until the selected information block is reached. With this system, access times of 20 milliseconds are possible. A multi-disc pack can store about 700 million bytes.

The popular form of disc is the *floppy disc,* a very flimsy disc of magnetic material contained in a protective sheath. The

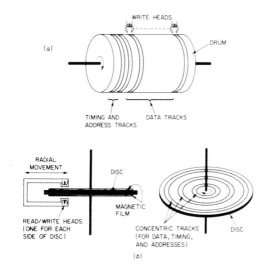

*Fig. 8.10. Two common types of magnetic store. (a)
Drum store. (b) Disc store*

complete unit is inserted in the disc-reading equipment. A development of this is the *stringy floppy* which is more like the cassette tape in appearance. It consists of a cassette or 'wafer' containing a *continuous* loop of narrow-gauge magnetic film. Using a special reader, this can be run at high speed, continuously. In this way it attains some of the recirculating advantages of high-speed access associated with the disc.

Bubble memory

So far, we have considered the principal types of data storage. RAMs allow rapid access but data is lost when the power is switched off. In addition the i.c.s required take up appreciable space in the computer. Magnetic storage can be permanent and is very compact, but it takes a relatively long time to transfer data between the store and computer. Bubble memories are a compromise. They are four times more compact then RAM i.c.s, yet allow access 100 times faster than magnetic stores. Since they contain no moving parts they are more reliable then mechanical recording systems. Data is permanently stored even when power is disconnected.

A thin film of magnetic material naturally becomes divided into domains, each magnetised in opposite directions. In each domain the molecules are all aligned in the same direction. Fig. 8.11 shows how the randomly-arranged domains are affected by an increasing magnetic field. One domain gradually decreases in

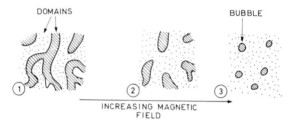

Fig. 8.11. Formation of magnetic bubbles

extent as more and more of its molecules become aligned in the direction of the field. Eventually only small islands or *bubbles* of this domain are left. These bubbles are the units by which we can store binary information: 'no bubble' = 0; bubble = '1.' The bubbles are only 1 to 5μm in diameter, so high-density storage of information is possible. Tens or even hundreds of kilobytes can be stored in a single d.i.l. package. Since the bubbles are like small magnets, they can be moved around by a rotating magnetic field (Fig. 8.12). Of course the actual molecules in a bubble are not moved: it is the *region* of magnetization that is relocated. Minute nickel-iron bar deposited on film become bars magnets as the field rotates and, as their magnetism changes, the bubbles are carried along. Fig. 8.13 shows a layout of a bubble memory. Bubbles (or 'no-bubble' gaps) are stored in minor loops. They can be transferred between the minor loops and the major loop by special magnetic transfer gates. They can circulate in the major loop to the detector, where the presence or absence of

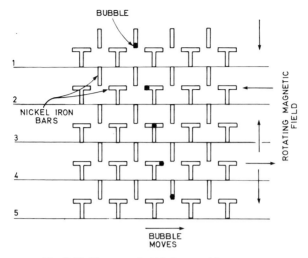

Fig. 8.12. The way a bubble is moved in memory

bubbles is read, and the result transmitted to the computer circuit. Bubbles can also be generated in response to incoming data. In addition to the memory chip, the memory device requires a small permanent magnet to provide the biasing magnetic field and crossed pair of coils to provide the rotating field needed for moving the bubble. The whole assembly is contained in an ordinary d.i.l. package.

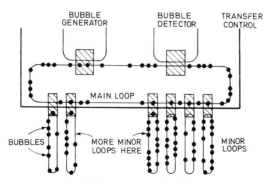

Fig. 8.13. Layout of a bubble memory

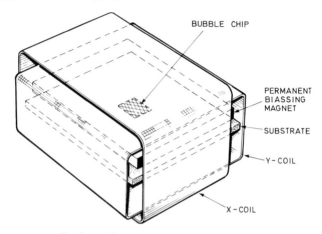

Fig. 8.14. The heart of the bubble memory i.c.

Hardware and software

When talking about a computer system, engineers divide it into two general parts—hardware and software.

Hardware refers to the physical units, such as the arithmetic unit, that make up the computer.

Software, on the other hand, is the term used for the programs that control the operation of the computer through the control unit, and enable calculations, fault detection, and so on to be performed automatically. There are many different types of program, and they are usually kept in a software library stored on magnetic tape, disc, or drum stores.

The program actually in use at any one time may be read from the library store by the control unit and stored in the fast access store to speed up operation.

The program contained in ROM is software in the sense that it is a program, yet is hardware in the sense that it is a fixed part of the computer. Such a program is often called *firmware*.

Types of program

Programs can be divided into a number of groups according to the type of operation they control. Each group contains a large number of more specific programs. A few important groups are mentioned here.

The basic programs which are used to perform large numbers of routine calculations (addition, multiplication, etc.) are known as subroutines, or housekeeping routings. They are not independent programs, but are incorporated by the programmer into his own program, as required. For example, a program may require an integration to be performed at some stage, and so the programmer refers the program to the appropriate integration subroutine rather than writing the integration routine into his program.

When an engineer writes a computer program he does so in one of a large number of different symbolic languages which have been devised to simplify his task. The computer cannot work directly in these programming languages and so the instructions have to be converted in to a machine language or code (a series of 0s and 1s). This is performed automatically by special programs known as assemblers or compilers.

Utility programs perform basic data-handling operations such as the transfer of data into a punched card file on to magnetic tape, or the assembly of data into a chosen sequence.

When writing a complex program it is necessary to check it out (usually referred to as program debugging) to make sure that is does the job required. Debugging is made easier by monitor and test programs that check out and locate errors in other programs. These monitor routines initiate print-out of the contents of the store at various stages of a program under test, and observe the progress of the program.

Program languages

Versatile and powerful as modern computers are, they cannot yet interpret instructions written in everyday languages such as English. It might be possible to build a computer that would understand instructions given in English but it would be very large and the cost would be enormous. Also, there are major problems to be overcome because many statements in English are ambiguous and require a knowledge of the context to be interpreted correctly.

Because of this, computer manufacturers and programmers have devised a number of specialized languages that are logically arranged to specify exactly and without ambiguity the operations that the computer should carry out. Writing of programs using these languages is a skilled task for trained computer programmers.

Many computer languages are symbolic so that the computer can interpret them simply without the need for complex assembler programs. However, it is common to write programs in clear language, that is, using simplified statements in a rather stylized version of English. This has the advantage that people concerned with the program but who are not trained programmers, can understand the various steps in the program.

The most popular language of this type is BASIC. Another English-based computer language is COBOL, specially designed for business use and probably the most widely used of all languages. ALGOL and FORTRAN are of special value in mathematical work, while PILOT is of particular use in computer-assisted learning programs. Each language is suited to its purposes in particular ways and probably no language can cover fully all the diverse tasks that present-day computers are able to perform.

9 Microwaves

The present advanced state of electronics is due largely to the impetus received from the development of radar during the second world war. The idea of radar was first thought of in the 1930s by a group of British scientists doing work on the propagation of radio waves. It was discovered during experiments that radio interference was caused by aircraft flying over the research station and on further investigation it was found that this was caused by radio waves being bounced off the aircraft and returned along the original path of transmission to the receivers on the ground. Development of this principle produced the first early warning radar equipment for detecting the presence of enemy aircraft.

In the early days, radar systems operated at frequencies of the order of 50 megahertz and conventional circuitry was used, but subsequently most radar equipment was designed to operate at frequencies between 500 and 15 000 MHz, in the so-called microwave region. This was made possible by the development of the magnetron, an oscillator valve which can produce high peak powers at this frequency. The microwave signals are fed to and taken from the aerial by means of waveguides instead of the transmission lines used in normal radio practice. The use of microwaves simplifies the aerial system, making it possible to design aerials that give a sharper, more clearly defined beam which aids in discriminating between near-by targets.

The principle of radar

The oscillator which is the heart of a radar set is pulse controlled so that it oscillates for only a very short period, often about one microsecond. These bursts of oscillation are used as the radar pulses. The pulses may be produced at intervals of one millisecond, that is a thousand pulses per second, and may be of relatively large peak powers–perhaps several megawatts in modern equipment–although the average power is low, perhaps only a few kilowatts.

The pulses of energy are fed to a highly directional aerial system and are transmitted along a path which may be almost as narrow as a searchlight beam. Immediately before the pulse is transmitted by the aerial the display system (which displays the information obtained by the radar installation), a cathode-ray tube, begins an 'X' trace (see Chapter 7). A fraction of the transmission pulse is fed to the receiver and appears on the cathode-ray tube as a small 'Y' deflection. This is called the *ground wave* and acts as a marker or reference for gauging the range of the echo or returned energy.

The transmitted pulse travels out into space and if it encounters a target, for example an aircraft or missile, or even in some

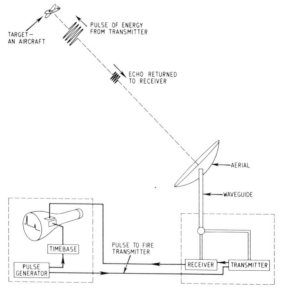

Fig. 9.1. Simple radar system

instances a rain cloud, some of the energy is reflected back along the original path and is received by the aerial system. From the aerial it is fed into the receiving system and after detection and amplification is applied to the 'Y' plates of the cathode-ray tube, lifting the 'X' trace as did the ground wave. As the speed of electromagnetic energy in free space is known, the distance of the target from the aerial system can be allocated. Also, as the speed of the display tube 'X' trace is also known, the screen of the display cathode-ray tube can be calibrated in miles or metres. The distance of the target from the radar set can thus be read off directly. As the energy has to travel to the target and back the distance travelled must be divided in half to get the true range when calibrating the range scale. Fig. 9.1 shows a simple radar system.

The Magnetron

The magnetron valve is the preferred source of high power oscillations in the microwave region. Recent years have seen many improvements in magnetron design but even now they are limited to mean powers of one or two kilowatts although the peak powers are measured in megawatts, enabling radar equipment to have a range of several hundred miles.

The action of the magnetron depends on the behaviour of electrons in crossed electric and magnetic fields. The d.c. electric field in the magnetron is supplied by the anode-cathode voltage of a thyratron valve during its conduction period. The magnetic field is constant, being supplied by a large permanent magnet situated, as shown in Fig. 9.2 outside the magnetron. When working on a magnetron care should be taken not to use heavy tools such as spanners and large screwdrivers made of ferrous materials: if they suddenly strike the body of the magnet, due to the attraction of the large magnetic field, they may damage the magnetron or distort the magnetic field. The engineer making any adjustment should also remove his wristwatch which may otherwise be permanently affected.

Under the influence of the magnetic and electric fields the electrons emitted by the cathode describe circular paths towards the anode. Some of the electrons move in a tight circle and return to the cathode but others eventually reach the anode, which is a block of copper with a series of cavities cut into it, the mouths of the cavities, as shown in Fig. 9.2, facing inwards towards the cathode. Movement of the electrons across the

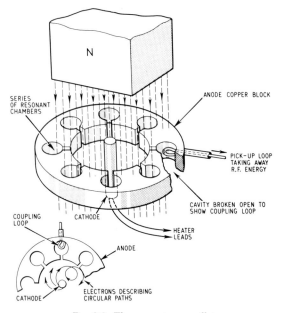

Fig. 9.2. The magnetron oscillator

mouths of these cavities sets up oscillations within the cavities and these oscillations in turn effect the movement of the electrons on their circular paths, producing a bunching effect. In this way energy exchanges occur between the cavities and the electron streams, the cavities being kept in oscillation during the application of the anode voltage pulse which provides the electric field. The frequency of the oscillations is determined by the physical size of the cavities–which in effect are tuned circuits and may be regarded as consisting of inductance and capacitance

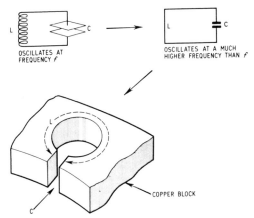

Fig. 9.3. Development of a cavity from an L, C circuit

as illustrated in Fig. 9.3. In practice the cavities are strapped together in a particular arrangement so that the oscillations reinforce each other and the magnetron operates in its correct mode. Power is taken out of the magnetron to the waveguide by means of a loop of cooper wire inserted in one of the cavities. This method of picking up r.f. energy, using a loop of wire or sometimes just a straight piece of wire sticking up in the r.f. field, is commonly used in microwave practice.

Waveguide system

In the early part of the nineteenth century scientists discovered that it was possible to send electromagnetic energy along a pipe of suitable dimensions. Although it is usual to think of an electric current as a number of electrons flowing along a wire, it is possible from a mathematical point of view to think of a magnetic and electric field configuration being guided along a path by a pair of wires. If this is accepted, the idea of a waveguide does not seem so strange.

If certain conditions for the two fields, electric and magnetic, are satisfied it is possible to make them move along a guide of rectangular section. The conditions for a mode commonly used, the H_{10} mode, are that the electric field is prependicular to the walls of the guide and the magnetic field is tangential to the walls, as shown in Fig. 9.4. Under these conditions copper walls may be fitted to the guide at specified distances apart so that the

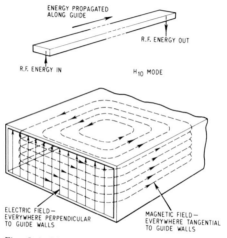

Fig. 9.4. Magnetic and electric fields inside a waveguide

guide becomes a rectangular section of any desired length. If a probe energized by, say, a magnetron is inserted in one end of this arrangement then some distance away, perhaps thirty feet or more, almost the same amount of energy can be removed with another probe. This is proof that the energy has been successfully guided along the waveguide.

At very high frequencies (above 10 000 MHz) energy loss in a waveguide system is less than it is for any other methods of transferring energy. Another advantage is that the mechanics of the aerial system are aided by the simple mechanical structure of the guide.

Aerials used in radar

In the microwave region of the frequency spectrum it is possible to use aerials which are small physically and which may have a variety of shapes. The shape of the energy beam that emerges from the aerial is determined by the shape of the aerial; many different kinds of aerial are in use according to the requirements of the system.

The effectiveness of an aerial is judged by its polar reponse, that is the extent to which it concentrates the transmitted energy in a beam. A diagram can be plotted to show this, such a diagram being called a polar diagram. This diagram shows the manner in which the field strength created by the transmitter varies with direction at a given distance from the transmitter's aerial. Fig. 9.5 shows a polar diagram of a cheese-dish aerial for two planes, vertical and horizontal. The wider the aerial dimension in any

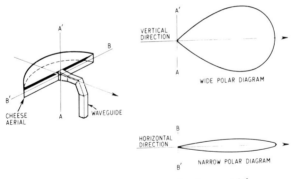

Fig. 9.5. Polar diagrams of a cheese aerial

particular plane, the narrower the polar response in the plane. This type of aerial is widely used on ships for detecting other surface craft, say during a fog. The fine discrimination needed to determine accurately the position of other vessels is achieved by the narrow horizontal polar reponse, the wide vertical response being of no significance.

Display of the signal

In Fig. 9.6 a typical 'A' scan display is shown. The ground wave
is obtained from the intial transmitter pulse and used as a marker
pulse. Returned energy after detection and amplification
appears on the cathode-ray tube as a further 'blip.' Since the

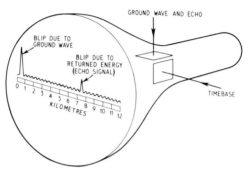

Fig. 9.6. Typical 'A' type scan

speed of radio waves in free space is known and the speed of the
cathode-ray tube 'X' trace can be pre-set, the real distance
between the transmitter and the target can be inscribed directly
on to the display tube face.

Other types of display include the plan position indicator
(P.P.I.). This type of display uses a special type of cathode-ray
tube. Firstly it has high persistence–that is, any brightening of
the trace remains for some time, perhaps many seconds, after
the brightening signal has disappeared. Secondly, a type of
sawtooth scan sweeps the spot repeatedly from centre to edge of
screen. This 'spoke' rotates slowly around the centre, synchronised with the rotation of the aerial. With a P.P.I., if there is no
echo signal nothing appears on the tube except a small bright
spot in the centre to indicate the transmitter position. When an
echo is received the scanning beam, which is rotating in synchronism with the aerial system, is suddenly brightened by
applying the echo signal to the grid of the cathode-ray tube
(instead of to the 'Y' plates as in the 'A'-scan). Thus not only the

distance but the position of the target is shown directly on the display. The trace for a P.P.I. display tube is produced by special coils fitted around the neck of the tubes, as in a television receiver, there being no 'X' and 'Y' plates. A typical P.P.I. display is shown in Fig. 9.7.

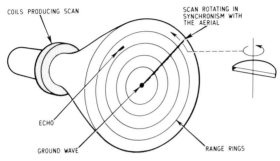

Fig. 9.7. A typical P.P.I. display

Some more recent radar equipments use a display system which only shows moving targets. This arrangement is useful in harbours and similar situations where it indicates only something which is actually moving, so that much useless information is excluded from the screen.

Doppler systems

In these, a beam of microwaves is sent out from a transmitter and the reflecting beam is received. If the beam has been reflected from a moving object, its frequency will have been altered. This is due to the Doppler effect, which in everyday life we notice when a fire engine rushes by sounding its syren. If the object from which the microwaves are reflected is approaching the transmitter, the frequency of the reflected waves is increased. The increase is very slight. For example a person walking toward the transmitter at 6 km/h causes the frequency of a 10GHz microwave beam to increase only to 10 000 000 056 GHz. This may seem a *relatively* small change, but actually it is an increase of 56Hz. If the signal produced by the reflected beam is mixed with a portion of the original output from the oscillator in the transmitter, the two signals interfere to produce a signal that has a frequency equal to the *difference* of their frequencies. This is the phenomenon known as *beats*. If you listen to two similar sounds that are roughly equal in loudness, but differ *slightly* in pitch you also hear a sound that throbs or beats at low frequency. This effect may be noticed when a propellor-driven aeroplane with two engines flies overhead and the engines are not running at exactly the same speed. In our example, the beat frequency detected in the receiver is 56Hz. This signal is detected and then amplified by an audio-frequency amplifier. Its frequency can be measured by electronic means and this tells us the speed with which the person is approaching the transmitter. This is the basis of the so-called 'radar' devices used by police to measure the speed of vehicles, using equipment placed by the roadside. It has many other uses for measuring velocity apart from this one. A simpler form of the device can be used to detect *any* slight movement of objects or persons within the range of the beam. Automatic doors in super-stores and airports use a microwave doppler device to sense when people are approaching the door. A couple standing in conversation by the door do not cause it to open. It is also used in intruder-detecting systems.

Microwave ovens

These use a magnetron and waveguide to produce a beam of microwaves (Fig. 9.8). The beam strikes a rotating fan or *stirrer*, the blades of which are designed to scatter the microwaves evenly throughout the oven, The waves are also reflected from the sides of the oven. If food (or similar material with high water content) is placed in the oven, microwaves can penetrate it to a depth of several centimetres. Water molecules in the food absorb the microwaves. In this way energy is transferred to the food and it becomes cooked. Since the microwave energy penetrates the food deeply, food cooks quickly in a microwave oven. By contrast with an ordinary oven which cooks by infra-red radiation being absorbed *at the surface* of the food, a

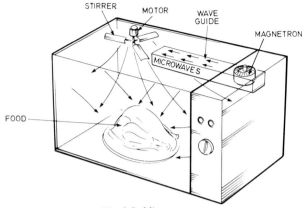

Fig. 9.8. Microwave oven

microwave oven cooks the food evenly almost all the way through. There is no outer roasted or crusty region. This is a disadvantage in cooking joints but a great advantage in heating up previously-cooked or deep-frozen foods.

Microwaves from space

One of the latest schemes to help provide the world with more energy is one proposed by NASA to construct dozens of geostationary Earth satellites each carrying several square kilometres of solar cells. These would absorb sunlight, converting its energy to electrical energy. This would be used to generate a beam of microwaves that would be directed toward an area on the ground (about 78 km^2) covered with receiving aerials. The energy of the received microwaves would be converted into electrical power at the rate of about 5 gigawatts per receiving station. This would be fed directly into national power networks. Such a scheme may seem far-fetched, and there are certainly some potential hazards to be overcome, but the technology required for it is already available. It is a distinct possibility that, given the will and resources, some scheme of this kind will one day come into being.

10 Medical Electronics

Electronics today plays an important role in medicine. This would be evident to anyone visiting one of the large teaching hopsitals where up-to-date techiniques in medicine and surgery are used and where large research projects are in hand. The co-operation between the medical profession and the engineering profession is unusually close and joint meetings are held regularly on electronic methods in medicine. In America university courses are held to teach the engineer something about medicine and the medical man something of electronics. Hardly a patient entering hospital for examination today will be exempt from the attention of some electronic apparatus.

What electronics can do in medicine

One of the most successful electronic aids to medicine is the EMI body scanner for which its inventor, Godfrey Hounsfield and Allan Cormack, shared the Nobel Prize for Physiology and Medicine in 1979, (Fig. 10.1). The scanner consists of a source of a narrow beam of X-rays and an X-ray detector that are opposite to each other and move round a circular path. The body of the patient lies at the centre of the circle. The detector measures what fraction of the X-ray beam is absorbed and feeds this information to a computer. As the apparatus rotates, the 'slice' of the body that is being scanned is penetrated by X-rays from all directions. The computer stores the information it receives. Parts that are in the shadow of bone when the beam comes from one direction are clear of shadows when the beam comes from

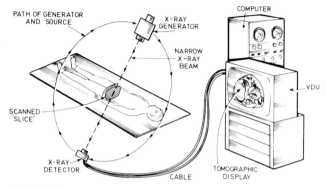

Fig. 10.1. Computer assisted tomography; the principle of the EMI whole body scanner

another direction. The computer is programmed to assemble this information and produce a display that shows the positions and shape of the organs and structures present in the slice. The earlier version of the EMI scanner were smaller and were used only for scanning the brain. They were successful in discovering brain cysts and tumours in scans lasting only a few minutes. Until then, it had been impossible to locate diseased brain tissue so precisely and so rapidly.

Measuring human electricity

Electronic equipment is also used to measure the small physiological voltages generated by muscle and brain tissue. These voltages are of importance in diagnosing conditions of the heart and abnormalities of brain function. The heart is a powerful muscle and in functioning it generates voltages which radiate throughout the body. The doctor can measure and record these signals by means of electrodes connected to the patient's limbs. These voltages form a pattern that is the same in all normal people so that divergencies from this pattern indicate some form of abnormality.

Fig. 10.2 shows a typical recording and the measurement being made on a patient, the technique being called electrocardiography (*e.c.g.*) The signals generated by the heart are picked up by silver or platinum electrodes attached to the wrists and the left leg of the subject. Measurement is made in turn of the

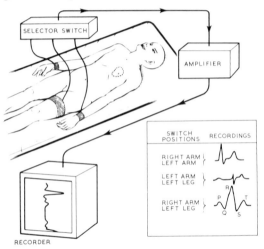

Fig. 10.2. Electrocardiograph equipment

variations in potential at the right and left arm, right arm and left leg and finally left arm and left leg. The voltages picked up in this way are fed to a suitable amplifier called a physiological amplifier. Since the human body is a high-impedance source of potential, we require an amplifier with FET input (p.141).

The output of the physiological amplifier is applied to a pen recorder which makes a permanent record of the signals from the patient–an electrocardiograph. The usual type of pen recorder cuts off at about 100 Hz and as this is too low for the adequate presentation of *e.c.g.* recordings special recorders are used which operate successfully up to 1000 Hz.

There are today many portable *e.c.g.* machines available which can easily be taken to patients in their homes for the examination of heart conditions.

Voltages from the brain

The investigation of brain functions can be undertaken with an equipment called the electroencephalograph (e.e.g.). The technique used is to measure the random voltages generated in the millions of nerve cells of the brain. The amplifiers used with this equipment need to have higher gains than those used in e.c.g. since the input signals from the brain may only be about 100 microvolts.

The voltages are detected at the patient's scalp by using small disc electrodes firmly attached to the skin by a harness. It is usual to use up to twenty-four electrodes. Each provides a series of signals which are fed to separate amplifiers and recorders. These amplifiers and recorders are built into a single complete unit as shown in Fig. 10.3. To save expense there are generally about sixteen amplifying channels and a switch is provided so that the physician can select various areas of the brain for investigation. The output of the selector is led to a pre-amplifier, one for each channel, and then to the main amplifiers and recorders.

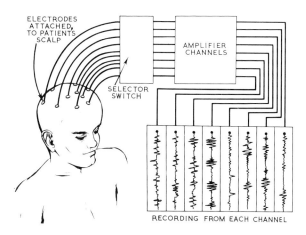

Fig. 10.3. Electroencephalograph equipment

Stimulators

The use of electronics to stimulate reactions in patients is perhaps most used in the field of cardiac medicine. The heart beats as the result of a nervous impulse which is conducted through the entire heart, beginning in the auricles and ending in the ventricles. This nervous impulse is electrical and can be imitated by electronic means. It should not be confused with potentials measured in e.c.g., the latter being the *result* of the heart beating. In some diseases a temporary or permanent disorder of the conducting mechanism occurs so that the heart beats too slowly or irregularly. Stimulators may in these circumstances be applied either externally on the chest wall or internally to the actual heart muscle. The stimulator produces a spike or similar waveform at regular intervals to make the heart beat with a regular rhythm. In chronic disease of this type a small semiconductor stimulator or 'pacemaker' is inserted in the body by surgical operation. It lies among the muscles on the region of the shoulder. Leads from the pacemaker pass to electrodes inserted in the heart muscles. Electrical spikes are delivered to the heart muscles so that they contract regularly and at the correct rate. The energy to power the pacemaker comes from batteries or from a small nuclear device in the pacemaker. This relies on the energy from radioactive plutonium-288. It can remain operating in the body for 10 years or more without further attention.

Another serious heart condition is heart flutter or fibrillation. In this condition, the heart works normally for most of the time but occasionally gets out of control and flutters wildly. Such a condition is fatal unless the normal beating of the heart is rapidly restored. The method of doing this is to administer a relatively large electric shock to the heart. This stops the heart for an instant, after which it generally resumes its normal action. When a person is affected, the shock may be given from a pulse-generator held against the chest. A new design, a miniaturised defibrillator, is permanently implanted in the body. It is connected to electrodes placed in the heart muscles. It monitors the normal beating of the heart and, if it detects the fluttering activity, it automatically delivers a shock to the heart muscle within a few seconds. Prompt correction of the heart's irregular action by this device has saved many lives.

Electronics helps the disabled

Persons who have been fitted with artificial limbs or who suffer from crippling diseases such as rheumatism often have difficulty in moving. If the doctor simply watches how the person moves, this does not usually provide enough information to allow difficulties to be corrected. In the Selpot system, the limb concerned has up to 20 infra-red LEDs fixed to it at points of interest, such as the joints. A pulse-generator worn by the patient flashes each of the LEDs in a regular sequence. The flashes are picked up by two infra-red cameras pointed at the patient, and signals are relayed to a computer. The computer is programmed to recognise the sequence of flashing of the LEDs, and is able to analyse the motion of the limbs and joints from this information. It produces a display showing the stages in motion of the limb and also calculates the forces acting at each joint. Such analysis is invaluable to the doctor in diagnosing the cause of the trouble and in prescribing remedies.

Radioactive tracers in medicine

It is useful to be able to follow the path taken by a chemical substance as it passes from one organ to another in the body. As an example, take a patient suffering from a disease of the thyroid gland, which is situated in the neck. We may want to be able o find out how well the gland is making its special hormone, thyroxine. The key element in thyroxine is iodine. We can give the patient food that contains iodine and see what the diseased gland does with it. Chemical analysis of tissues or body fluids does not give a complete picture of the process. Chemical tests do not allow us to distinguish between iodine recently supplied in food and iodine that has been in the body for a long time.

However, if the recently supplied iodine has mixed with it a minute quantity of *radioactive* iodine, we can see the radioactivity to distinguish this batch of iodine from all others. The radioactive iodine acts as a *tracer*. Furthermore, the radiation from radio iodine passes through the tissues and out of the body, so we can detect it from outside. There is no need to remove tissues of fluids from the body for testing. Radiation from radioactive substances causes ionization. When such radiation passes through a semiconductor it generates extra charge carriers. Solid-state radiation detectors work on a similar principle to the photodiode (p.185). Electron-hole pairs are generated and leakage current is increased. The most commonly used radiation detector is not a solid-state device. It is a discharge tube, called a Geiger-Müller tube (Fig. 10.4), and contains a gas under low pressure, usually argon. When ionizing radiation passes through the tube, ions are formed in the gas. The p.d. between the outer

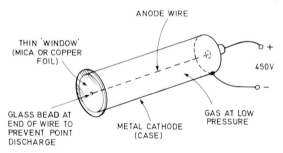

Fig. 10.4. A Geiger-Müller tube: The thin window admits α-particles or β-particles. X-rays can penetrate the case too

cylinder and the axial wire is such that the additional ions allow conduction to take place through the gas. Discharge occurs and the pulse of current is detected electronically. Pulses may be counted individually or the rate of occurence of pulses may be measured by a rate-meter circuit. When doctors use this equipment the patient is fed with food containing normal amounts of ordinary iodine plus a minute and harmless amounts of radio iodine. A Geiger-Müller tube is then placed against the patient's neck where the thyroid gland is situated. After a while the instrument shows that increasing amounts of radiation are coming from this region. By measuring the changes in the amounts of radiation, medical scientists are able to investigate the behaviour of the diseased gland.

Radioactive tracers are widely used in medicine, in scientific research and in industry. By their use we can detect and follow the movement of exceedingly small quantities of materials–often only a few micrograms. The techniques can detect minute leakages in drainage systems, or the wear in the bearings of car engines. The Geiger counter is used not only with trace elements, but in monitoring of the much larger levels of radioactivity encountered in nuclear power generation and processing plants. In this connection too it is important to the safeguarding of health, both of the workers and that of the general public.

Inside the stomach–the radio pill

Radio pills are, as their name suggests, devices to be swallowed. The idea is that they transmit information which can be detected, recorded and later assesed by the physician during their passage through the human alimentary system. Before the advent of the transistor it was impossible to make a device small enough to be swallowed without considerable discomfort to the patient. Now, by using a transistor circuit powered by a suitable cell, pills can be made which are less than 20 mm long and about 8 mm in diameter. Fig. 10.5 shows a typical radio pill which will transmit information about the pressure, temperature, acidity, and many other factors of interest to the physician. The pill is essentially a minute radio transmitter working at about 300–400 kHz and the information is radiated through the body to a sensitive receiver outside, which detects the signals and passes them to a recorder. Information is conveyed by varying the frequency of the transmitted signals (frequency modulated), and this can be done in several ways. If for example it is desired to measure the pressure inside the intestines, then the inductor in the circuit can be made to alter the tuning of the transmitter, its iron core being displaced in accordance with the pressure on the walls of the pill.

If the acidity of the stomach is to be measured this can be done by placing pH electrodes on the surface of the pill. When immersed in the acid solution in the stomach the electrodes generate a voltage proportional to the acidity, and this is used to change the frequency of the transmitter by altering the potential

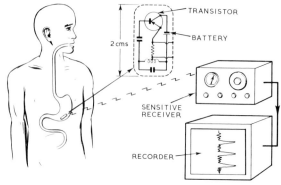

Fig. 10.5. A 'radio pill'

on the transistor. Again if the temperature is to be measured then the effect of temperature on the transistor can be used to vary the frequency.

Although the pill-transmitter can hardly be an accurate, close-tolerance device, the receiver should be as selective as possible allowing for the bandwidth required to receive the frequency changes from the radio pill. Typical radio pill transmitters may drift by as much as one per cent off their normal centre frequency so that the signal frequency variations should be at least ten per cent greater to give a good signal-to-noise ratio.

Ultrasonics

Ultrasonic techniques have a certain similarity to radar. Bursts of ultrasonic (i.e. frequencies above about 16 000 Hz) energy, pulsed as in radar, are transmitted and reflected and the returned echoes are received and can be plotted on a recorder.

This energy can be applied at certain points to the body, for example the skull and the abdomen. In the ultrasonic scanner the body of the patient is scanned by a narrow beam of ultra-sound. Part of the beam is reflected at the boundary layers between organs, where sudden changes in density occur. The reflected ultrasound is detected and information fed to a computer. This produces a 'picture' on the VDU, showing internal organs of the body. The technique provides information which can be used as a support to X-ray examination and other techniques to tell the physician more about his patient and to aid his diagnosis. The ultrasonic scanner is of particular use in examinations during pregnancy. The frequent use of X-rays in such circumstances could cause genetic damage whereas ultrasound is safe.

A further use of ultrasonics in medicine is the ultrasonic hypodermic needle. The needle incorporates a suitable transducer to which is applied the output of an ultrasonic generator. The ultrasonic transducer makes the needle act rather like a road drill. Its advantage is that it can be used for penetrating deep body tissues without damaging surrounding tissue or causing pain.

The sterilization of surgical instruments is another sphere where ultrasonic techniques are playing an important part. The ultrasonic transmitter (e.g. a piezo-electric transducer) is placed in the sterilization tank together with agents, for example detergent, to lower the surface tension of the water. When the transmitter is switched on the micro-agitation it produces in the water removes dirt from the instruments and makes subsequent sterilization much easier.

11 Radio and Television

In Chapter 2 it was explained how a high-frequency alternating current can generate the kind of electromagnetic radiation that we know as radio waves. The essential features of a radio transmitter are an *oscillator* to produce the high-frequency a.c., an *amplifier* to pick up sound or musical signals that are to be transmitted, and a *modulator* by which the information present in the amplifier signals may be super-imposed on the high-frequency a.c. The details of the circuits concerned are beyond the scope of this book, but the block diagram of Fig. 11.1 shows the main features. For further details the reader is referred to the companion *Beginner's Guides* to radio, to television and to colour television, all by Gordon J. King.

For radio transmission on long-wave, medium-wave and short-wave bands, the high-frequency oscillation, or *carrier wave,* usually carries the signal by means of amplitude modulation (A.M., Fig. 11.2). This method has been widely and successfuly used almost since the beginning of radio transmission. As its name implies, the signal is used to alter or modulate the amplitude of the carrier wave. The circuits required for transmission and reception of A.M. are relatively simple–the most basic of all radio receivers uses only a single diode as detector. The main disadvantages are that static electric discharges in the atmosphere and locally-generated interference, caused by the operation of switches and motors, can produce a considerable amount of unwanted noise at the receiver. One way around this is to use *frequency modulation* (F.M., Fig. 11.2). To be successful, F.M. must be used only on a very high frequency (V.H.F.) radio transmission, or at higher frequencies (Table 2.1, p.31).

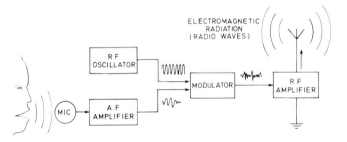

Fig. 11.1. Block diagram of radio transmitter

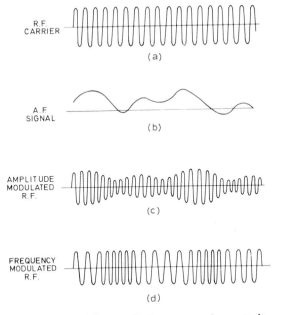

*Fig. 11.2. Modulating radio frequency carrier waves. In
(d) the effect is exaggerated for clarity*

Radio receiver

The main elements of a radio receiver are shown in the block
diagram of Fig. 11.3. The tuned circuit often consists of a
capacitor-inductor circuit. The capacitor is variable so that the
resonance frequency of the tuned circuit can be made equal to

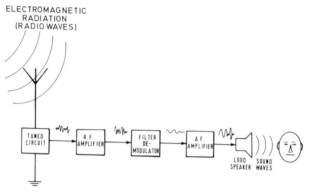

Fig. 11.3. Block diagram of a radio receiver

the frequency of the transmitting station. In this way, one station
may be received and stations operating on other frequencies will
not be received. Oscillations in the tuned circuit are tapped and
fed to an r.f. amplifier. Following this the signal is demodulated,
removing the r.f. component and leaving only the original audio
signal. This may be amplified further before being fed to a
loudspeaker. In a F.M. set, demodulation consists in comparing
the incoming radio signal with a signal generated in the set and
having the same central frequency as the transmitter.

   Television is the transmission and reception of moving pictures over a distance. Like the cinema, television makes use of
the persistence of vision: if the information to form the pictures
can be assembled sufficiently rapidly on the screen being viewed,
the viewer will see a complete picture.

   When light rays enter the eye and strike the retina at the back
of the eye the impression they make does not cease immediately

but remains for some time afterwards. This persistence of vision is long enough for separate pictures, following one another in time, to give the impression of continuous movement. Each picture follows its predecessor rapidly enough for the eye still to be occupied with the previous one, which it replaces. The impression of movement is thus obtained from a series of still pictures. In the television studio the scene to be transmitted is first 'photographed' at intervals by the television camera.

Scanning

Before transmission each of these still pictures must be reduced to a series of electrical signals. This is done by scanning the picture. The scene, by being scanned from left to right in a series of lines, is broken up into a series of light impressions of varying intensity. These are reassembled at the receiver to form the picture seen by the viewer.

The camera

Fig. 11.4 shows a simplified form of the image orthicon tube. The scene to be televised is focused on to the photocathode by a normal optical system. Under the influence of the light from the scene the photocathode emits electrons in quantities which depend on the amount of light falling upon it–those parts of the scene which are bright cause more electrons to be emitted than those which are dark. These electrons are attracted to the target where in this way an electronic image of the scene is produced, the image changing continuously as the original scene changes. At the other end of the tube an electron gun emits an electron beam which scans the target, rather as the eye scans the page of a book. In doing so it samples each part of the electronic image. Some of the electrons of the beam return, forming a return beam, others remaining on the target (to neutralize the electrons given off by the photocathode). The return beam is thus varied in strength, depending on the electrical value of the electronic

image at each point. For instance, a strong light on the photocathode produces a large electric charge on the equivalent part
of the target and when the beam samples this point a large
change is produced in the return beam. The return beam passes
through an image amplifier which increases the strength of the

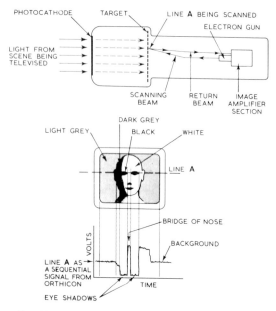

*Fig. 11.4. Principle of operation of the image orthicon
television camera tube and the way in which a scene is
scanned to provide a sequential electrical signal*

signal by up to 2000 times. The camera tube thus provides a
sequential series of electrical signals of value equal to the
brightness of the various parts of the picture.

The speed at which the picture is scanned is such that a
complete picture is produced every 1/25th of a second. This is
sufficient to give the viewer the impression of watching a
continuous event.

For colour television the image orthicon camera has been replaced by the vidicon camera tube, which is described on p.208. Before describing the additional complexity of the transmission of colour pictures, we will look at the general principles of television transmission and reception.

Interlacing

To enable the electron image on the target of the image orthicon to be scanned the scanning beam must be deflected–both horizontally and vertically. Coils are wound round the tube and by applying suitable waveforms to them varying fields are produced in the tube to deflect the electron beam. These waveforms are provided by the timebase circuits, the frame (or field) timebase deflecting the beam vertically and the line timebase deflecting it horizontally. In the television receiver the cathode-ray tube is deflected in exactly the same way to trace out pictures, 25 frames per second in most systems. It has been found that if the frames are split so that half the lines are scanned in 1/50th second and then the beam returns to scan the alternate lines during the next 1/50th second, then although a complete frame is still scanned only once every 1/25th of a second the eye will be given the impression of seeing a frame every 1/50th of a second and flicker is reduced to negligible proportions. This technique is called interlacing.

The synchronizing pulse generator

In the television receiver and in the camera, electrical signals are needed so that the deflections of the electron beams in the camera and the receiver cathode-ray tube are synchronized. These signals are provided by the synchronizing pulse generator, which is shown connected in the system in Fig. 11.5.

The sync. generator, as it is called, consists of a 'master-oscillator' and a number of counting-down, gating and mixing circuits. It has four outputs: (1) A series of pulses which are fed to the camera tube line timebase to control the horizontal scanning of the image orthicon target in a regular manner. These are called the line sync. pulses (2) A series of pulses which act in the same way to control the vertical scanning of the camera tube target. These are called the field or frame sync. pulses. Both these pulses are a little ahead in time of the other pulses from the generator–the camera tube has to start operating before the rest

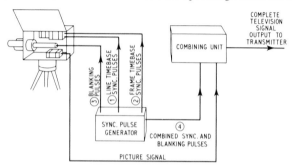

Fig. 11.5. Sync pulse generator used at the studio

of the equipment. (3) A series of blanking pulses. These are fed to one of the camera tube electrodes and virtually switch the tube off during the time when the electron beam is flying back to the start of a line or frame, in order to suppress the flyback traces. (4) Combined sync. and blanking pulses. This output is fed to a mixer unit where it is combined with the picture information to make up the composite television waveform. The composite series of signals is then fed to the transmitter with the scanning information included in it to enable the domestic receiver to scan in synchronism with the camera tube in the studio.

The composite television signal, called the video signal, may be connected to Telecom lines, or in some cases, to a radio link, before being fed to the transmitting station, which may be many miles away from the studio. The sound signal is separate at this stage, being obtained by using the same techniques as for ordinary sound broadcasting–microphone, amplifiers and so on–and fed to the transmitter over a separate line.

The transmitter

Transmitting stations are usually located as high above sea level as possible and also central to the area they are to serve. This often means that they are situated in rather remote places. In the station, apart from the transmitter itself there is control gear and monitoring equipment. In practice there are two transmitters, in case of a breakdown. A local transmitter may have a power output of the order of 5 kilowatts. The video and sound signals from the studios are sent to the transmitting station and then fed to the modulating stages of the transmitter. The video signal is transmitted by A.M. and the sound signal is transmitted by F.M.

The aerial system consists of an array of dipole elements connected together to give either a directional or an all-round, omni-directional transmission depending upon the location of the transmitter and the area to be served.

The combined signal is received on an aerial array that is usually of the type shown in Fig. 11.6. This is a polarized Yagi array, for the reception of u.h.f. signals that are horizontally polarized. The aerial may also be mounted with its elements vertical for reception from aerials that transmit vertically-polarized waves.

The aerial in the figure has two coupled dipoles for the reception of the signal. These are connected to the receiver by a coaxial cable. Behind the dipoles is a metal grid, acting as a

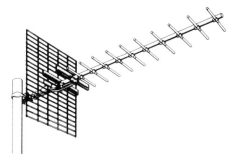

Fig. 11.6. A practical aerial array for colour television reception

reflector. Radio waves incident on this are reflected back towards the dipoles. This assists in the reception of weak signals, a factor that is important with u.h.f. transmissions as they weaken appreciably with distance. The elements in front of the dipoles are metal rods called *directors*. These help make the aerial array more directional. Signals arriving from the direction in which the array is pointing are strongly received while signals coming from the side or rear are received only weakly. This helps in picking out the transmission from the selected broadcasting station against the background of electrical noise from other sources and the transmissions of other stations within a receivable radius. Improving the signal-to noise ratio in this way helps considerably in obtaining a good television picture.

The receiver

The signal from the aerial is fed into the receiver by means of a coaxial cable. This cable is specially constructed to match the aerial system to the receiver input and also to screen the signal from stray voltages which may otherwise be picked up by the down-lead. The metal sheath around the coaxial cable provides this screening.

Fig. 11.7 shows the basic television receiver in block form. The first stage is an amplifier capable of handling both the vision and

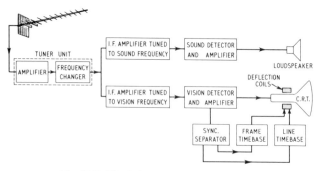

Fig. 11.7. Block diagram of a television receiver

sound signals. This amplifier is incorporated in the tuner, which enables the correct channel selection to be effected. The initial r.f. amplifier stage is followed by a frequency changer stage which changes the carrier frequency of the selected and now amplified channel to a much lower, fixed frequency called the *intermediate frequency*. This is always the same whichever channel is chosen. Reducing the frequency simplifies the design of the following i.f. amplifiers needed to amplify further the signal. The output of the i.f. amplifiers is fed to the sound and the video detectors. Tuned circuits are incorporated in the i.f. stages to separate the sound and vision signals. The detectors remove the carrier wave, leaving just the sound and vision signals which are applied to suitable output stages for final amplification. The sound output stage drives a loudspeaker. The vision output stage feeds the picture tube and also the sync. circuit where the sync. pulses are removed from the composite waveform. The sync. pulses are then used to control the timebase circuits which, in turn, control the scanning of the cathode-ray tube. The timebase circuits control the *movement* of the electron beam in the cathode-ray tube, the output of the video output stage *modulating* the beam so as to control the *brightness* of the spot it produces.

Colour television

It is not possible to transmit colour as such: it is only possible to transmit a code. For black and white transmission this code is one of amplitude. White signals are represented by high amplitude video signals. In the case of colour, cameras fitted with red, green and primary colour filters take out from the scene those electrical impressions that correspond to that particular colour. Remember that a camera is merely a device for converting light variations into electrical current variations.

These signals are coded into what are called luminance and chroma signals and then used to modulate a u.h.f. carrier (Fig. 11.8). At the receiving end they are detected, analysed, and amplified and then used to modulate one of the cathode-ray tube guns in the cathode-ray tube. Each gun corresponds to one of the colour filters at the studio, that is red, green or blue and so the picture is reassembled as a coloured picture.

Apart from the coding, the basic principles of colour television are the same as black and white television.

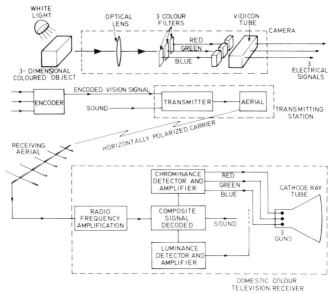

Fig. 11.8. Transmission and reception of colour television reduced to its simplest terms

The black and white u.h.f. receiver simply takes the luminance signal and produces a black and white picture while the colour u.h.f. receiver needs both the luminance and chroma signals to recreate the colour picture.

Apart from the luminance signals, with the high-frequency chroma signals superimposed on them, the colour television signal (Fig. 11.11) consists of a *line synchronising pulse* and a *colour burst*. These both occur between the transmission of the lines themselves. The 'line sync' pulse is required to synchronise the scanning of the receiver tube to that of the camera. On either side of the pulse the signal level is equivalent to that of black.

The colour burst consists of about 10 cycles of a high frequency sub-carrier wave. This is a signal used in the transmitter in producing the chroma signals. It is not transmitted except as the short colour burst. This burst is used in the receiver to synchronise a generator in the receiver that is producing an identical sub-carrier wave. With the generator in the receiver exactly synchronised to that in the transmitter, the chroma signals can then be demodulated accurately.

Fig. 11.8 shows the block diagram of a simplified colour television system from the studio to the transmitter and finally to the receiver. The audio side is excluded since this runs parallel and is easily understood. The camera with its three filters, called dichroic lenses, is focused on to a scene or object lit with strong white light. The three images, red, green and blue are images converted into electrical signals and encoded to give a combined luminance and chrominance vision signal which is sent to the transmitting station with the sound signal for transmission.

On reception the weak signal is amplified and decoded into chroma, luminance, and sound signals. The luminance and chroma signals are fed to a triple gun cathode-ray tube to create the final picture by individually exciting a matrix of red, green, or blue coloured dots on to which each gun is individually focused. The result, due to the persistence of human vision, is a naturally coloured picture.

The three-gun cathode-ray tube

In colour television the face of the tube has to be treated with three different phosphors to produce a colour picture.

The object of the three-gun tube is to vary the intensity of the beam from three separate electron guns so that each will carry the information appropriate to one of the primaries. Each beam is made to strike the appropriate primary colour phosphor dot, making it glow. Because of the close proximity of the three dots the eye is deceived into thinking that the three primary dots occur at the same point and so it adds them together to give a colour of the correct hue, saturation and luminance to correspond with the exact point in the original studio picture.

Operation

Looking at Figs. 11.9 and 11.10 we can see how the colour tube works. It is in fact three tubes in one glass envelope and using the same tube face. The three electron-gun assemblies work completely separately to produce three separate electron beams each of a different strength according to the signals they are carrying. Due to the geometry of the tube the three beams converge on a small group of holes in the shadow mask and pass through them to diverge on the other side. The distance between the shadow-mask and the colour-dot screen is accurately controlled so that the beams each fall on the correct colour dot–the red electron beam on the red dot, the green beam on the green dot and the blue beam on the blue dot.

Each dot is then activated by an electro-luminescence process to give out an amount of light proportional to the strength of the electron beam striking it. The three dots thus activated are so close physically that the eye is unaware of their separation and so adds the three primaries together to give the intended colour for that point.

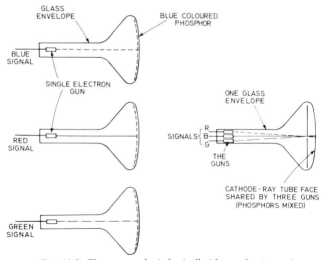

Fig. 11.9. Three-gun tube is basically 'three-tubes-in-one'

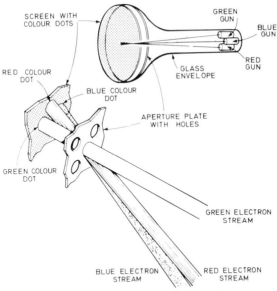

Fig. 11.10. Action of three-gun colour tube

The arrangement of the dots, their relationship to the shadow-mask and the arrangement of the holes in the shadowmask relative to the electron guns, demands a high degree of skill on the part of the tube manufacturer.

Arranging the colour dots

To produce the required pattern of different coloured phosphor dots a rather ingenious method is used. First a layer of potassium silicate is laid on the glass screen to prevent any reaction between the oxides in the glass and the phosphors. The green phosphor is then flow-coated on to the screen and dried. Now the shadowmask is fitted in place and a narrow beam of ultraviolet light is shone through the holes in the screen from a source in the exact position of the electron gun corresponding to the colour green. Where the light falls on the phosphor it hardens (in a manner similar to photo resist) forming a small dot in line with the source. The shadow mask is then removed and the screen washed to get rid of all the unexposed green phosphor, leaving a pattern of green phosphor dots on the screen. This process is then repeated in turn for the blue and red phosphors to cover the screen with groups of tiny green, blue and red dots.

It is important that the same shadowmask is used throughout the production and in the final tube so that the dots will line up

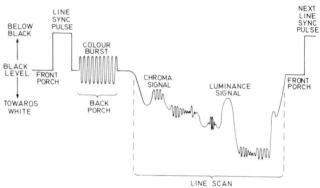

Fig. 11.11. Colour television signals (not to scale). The high frequency is superimposed on only part of the luminance signal in the diagram

correctly with the three electron guns irrespective of any small manufacturing differences. After finally clipping the shadow-mask into position, thin metal screening plates are fitted round the periphery to prevent stray electrons escaping round the edges. Such electrons would activate the phosphors and cause colour dilution.

One tube colour camera

Early cameras for colour television used three image orthicon tubes far too big for a spacecraft to carry. Today the vidicon one-tube camera is small enough and light enough for spacemen to use on the Moon shot.

The vidicon tube is shown in Fig. 11.12. As in the image orthicon tube, the electron beam is made to scan a target on

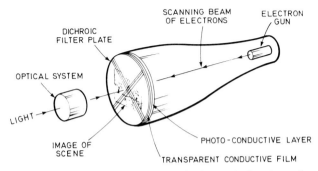

Fig. 11.12. Vidicon colour camera tube (magnetic focusing coils and field coils omitted). Beam control system omitted

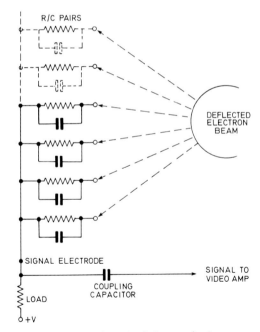

Fig. 11.13. Equipped circuit of photoconductive camera tube showing effective RC elements

which the image has been focused. The target consists of a photoconductive material, lead oxide (c.f. LDR, p.88) that has high resistance in the dark and low resistance in the light. The material of the target acts as a very large number of 'resistors' in parallel with 'capacitors,' as indicated in Fig. 11.13. The photoconductive material provides the resistance; the layer of photoconductive material and the transparent conductive layer provide the capacitance. As the electron beam scans the target, each of these elements is brought briefly into circuit by contact with the beam. During this time a p.d. builds up across the 'capacitors.' The beam then moves on, to complete its scan. In the meantime, charge leaks away from the 'capacitor' via the 'resistor'–the rate at which this occurs depends on the brightness of illumination at that part of the screen. In a brightly lit area, the 'capacitor' may discharge almost completely. In a dark area it may retain almost all its charge. When the electron beam contacts the element at the *next* scan, the amount of current that flows depends on how much charge has leaked away since the previous scan. Thus, as the beam scans each element in turn, a varying current passes through the load resistor to each 'capacitor' in turn. Variation in current causes variation at the coupling capacitor. This alternating voltage passes across the coupling capacitor and becomes the video signal.

In the three-tube camera, three pictures are taken from the screen by the use of dichroic filters (filters that transmit light of one colour and reflect light of all other colours). Each filter passes its red, green, or blue picture respectively to one of the tubes.

We thus have three signals — red, green, blue, which can be used to create the luminance signal necessary for black and white pictures.

Looking at this situation another way, if we have the luminance signal and two of the colour signals it is possible to derive the third missing colour signal necessary for colour pictures.

This is what takes place in the vidicon one-tube camera. Two sets of dichroic filter strips are created at 45 degrees, one that lets through red, and one that lets through blue light.

When the discharging electron beam scans the tube face it picks up three signals due to the pattern created by the striped pattern: the luminance signal, a red signal, and a blue signal. This is solely due to the ingenious pattern created by the dichroic filters and the geometry of the pattern. Thus from the three signals we can recreate the four signals we need; red, green, blue chroma and the luminance signal.

Other standards

The descriptions given in this book are applicable to most systems of colour TV. However, there are *three* main systems in use; PAL (used in UK); NTCS (used in USA); SECAM (used in France, USSR) etc. These systems differ in detail so that signals generated in one system cannot be received in another. Television programmes are sometimes originated in one country and are to be shown in another country that uses a different system. Some method of converting signals must be employed. The earliest method was to display the incoming programme on a compatible television set, then pick up the picture using a camera of the locally favoured system. Such a conversion produces loss of definition and colour quality. A more recent method is to modulate an ultrasonic signal with the demodulated television signal. The ultrasound is transmitted into a large quartz crystal; it is reflected several times at the faces of the crystal until it reaches another transducer, when it is converted back into an electrical signal. This can be transmitted according to the required standard. The path of the ultrasound is such that one complete television frame is stored as a beam of ultrasound in the picture. This works well, though echoes in the crystal cause some degradation of the image.

The third method involves digital electronics. In this method the incoming analogue signal is converted to digital form (see next chapter). Four complete frames are stored. Since the NTSC system uses 525 lines at 60 frames per second while PAL uses 625 lines at 50 frames per second, an exact frame-for-frame and line-for-line conversion can never be made. However, in the digital system the previous two frames and following two frames are compared and the contents of 'missing' frames or lines are calculated by computer and then transmitted. Such a converter shows no loss of picture quality.

The television set and the information era

Apart from its obvious uses in entertainment and education, television provides ready access to a variety of sources of information. One channel of information is *teletext,* in which pages of data are transmitted at the same time as normal programmes. In the 625-line system, only 575 lines are used for the picture. The remaining 50 lines, which do not appear on the screen, contain synchronising pulses for the beginning of each frame, test signals to allow for testing of receivers when Test Cards are not being shown, or they are used to allow the beam time to retrace its path to the top of the screen at the end of each scan. The spare lines also carry the teletext signals, with enough information to transmit four 'screenfuls' (or pages) of text a second. On BBC channels the teletext service is called *Ceefax,* on ITV channels it is called *Oracle.* The information presented includes news items, weather forecasts, and numerous other subjects of a 'magazine' nature. There is also a facility whereby subtitles may be superimposed on the ordinary television picture to aid deaf viewers. When a particular page of data is required, the viewer selects it by pressing numbered buttons on the teletext control panel. When the signals for this page is received it is stored in the memory in the teletext section of the television set. Then it is displayed on screen for as long as required.

Viewdata is a second system in which the television set receives data from a computer over the ordinary telephone system. The version of viewdata operated by British Telecom is known as *Prestel.*

A vast store of information (over 180 000 pages) can be accessed by Prestel–news, weather, sporting events, entertainment, job vacancies, shopping news, advertisements, transport time-tables and much more. There is also the 'response frame' which provides for message services and for interaction between viewer and computer. For example, having accessed a list of hotels, the viewer might then be able to arrange a booking through the computer, even paying for the room by typing in a credit card number!

12 Recording

The principle of recording sound by convolutions in a spiral track in the surface of a rotating disc was known and used widely long before electrical recording and reproduction were developed. Today we use piezo-electric or electro-magnetic transducers to produce or to detect the vibrations of the stylus, and the results are certainly superior to those of the older mechanical systems. The major contribution of electronics has been in the processing of signals during recording and reproduction. In particular, electronics makes it possible to combine two or more signals and record them as one. This allows sounds from different sections of an orchestra to be balanced so as to obtain the most pleasing effect. The mixing of signals is of importance in stereophonic recording. The advent of this technique in the nineteen-fifties gave an enormously increased realism to recorded music. Two microphones are used in recording, placed so that one receives sounds as they would be heard by the left ear of a person in the audience, while the other receives the sounds that would be heard by the right ear. These two signals are recorded in a single groove to carry the two tracks (Fig. 12.1.). It is arranged that if L and R signals are equal in amplitude *and in phase,* the coils are energised in opposite sense, causing the cutter to move from side to side. This condition could occur if an instrument were exactly centred on the 'stage' though this is unlikely, since echoes and other effects occur that almost always produce slightly different signals at each microphone. However if both coils are fed with the *same* signal, as in a monophonic recording, the head produces the same side-to-side modulations of the grove as are found on monophonic records. This gives compatibility between the two systems. Several types of cartridge are available for

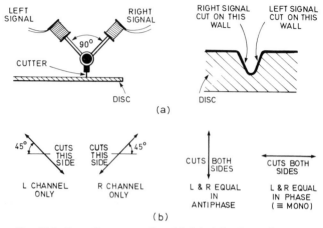

Fig. 12.1. Recording a stero disc. (a) Principle of recording two channels in one groove. (b) Motion of the cutter

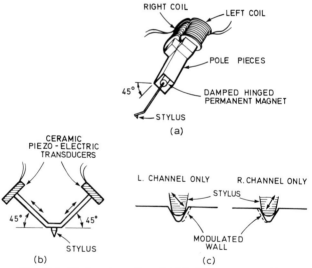

Fig. 12.2. Playing stereo discs. (a) Moving magnet cartridge. (b) Ceramic cartridge. (c) Motion of the stylus

playback (Fig. 12.2). In the moving-magnet cartridge the permanent magnet oscillates between two pairs of pole-pieces. Each pair carries a coil in which currents are induced as the magnetic field changes. The moving-coil cartridge operates on a similar principle. In the ceramic cartridge, two piezo-electric transducers are set at right angles to each other, one for each channel. The diagrams show modulations in one wall or the other, but in most recordings *both* walls are modulated, sometimes similarly, sometimes differently, depending on the signals carried.

A later development is quadraphonics, or 'surround sound,' in which three or more speakers are used to produce an even more realistic effect. This has never become popular, perhaps because the relatively small improvement in the effect is not worth the considerably increased cost of equipment, and perhaps because the manufacturers of rival systems never agreed to standardise their products.

Magnetic tape recording

The magnetic tape recorder became available in the late nineteen-forties and has increased in popularity ever since. The facility of being able to record at home without requiring additional or expensive equipment has been a big factor in its widespread use. Earlier forms of magnetic recording made use of wires or paper tape coated with magnetic materials. Today, coated plastic tape is used almost exclusively. The principle is illustrated in Fig. 12.3. The coating of ferric oxide or powdered iron can be thought of as a large number of minute regions known as *domains*. In any one domain, all the molecules of the material are arranged in one direction, so that the domain is in effect a small magnet. In unmagnetised tape the domains are magnetised in different directions (Fig. 12.3a) so that there is no magnetic field around the tape as whole. If the tape is fully magnetised (saturated), all domains are magnetised in the same direction (Fig. 12.3b). In a tape that carries a recording an intermediate state exists in which a proportion of domains are aligned, but the remainder point in random directions. The

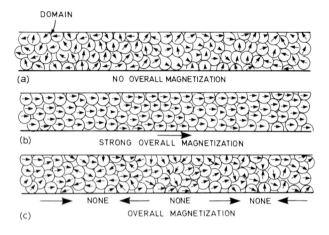

DOMAIN

(a)　　　　NO OVERALL MAGNETIZATION

(b)　　　STRONG OVERALL MAGNETIZATION

→ NONE ← NONE → NONE ←

(c)　　　OVERALL MAGNETIZATION

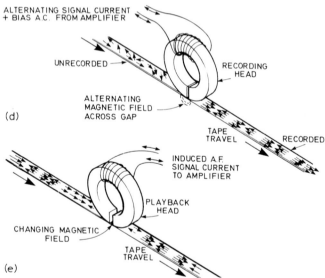

ALTERNATING SIGNAL CURRENT
+ BIAS A.C. FROM AMPLIFIER

UNRECORDED

RECORDING HEAD

ALTERNATING
MAGNETIC FIELD
ACROSS GAP

(d)

TAPE TRAVEL

RECORDED

INDUCED A.F.
SIGNAL CURRENT
TO AMPLIFIER

PLAYBACK HEAD

CHANGING MAGNETIC FIELD

TAPE TRAVEL

(e)

Fig. 12.3. Magnetization of tape. (a) Unrecorded tape. (b) Saturated tape. (c) Recorded tape. The size of the domains is very much exaggerated. Arrows represent direction of magnetization. (d) Recording. (e) Playback

proportions of domains similarly aligned depends on the strength of the magnetic field as the tape passes the recording head. On play-back, the strength of field around the tape, as detected by the play-back head, depends on the proportion of domains similarly aligned. Thus when the tape passes by the play-back head it reproduces the varying field that the recording head originally produced. Recording of sterophonic sound is easy with a tape recorder. It simply needs two heads, one for each channel and produces two parallel tracks on the tape.

When a region of tape is magnetised during recording, the speed of the tape must be such that this region has moved away

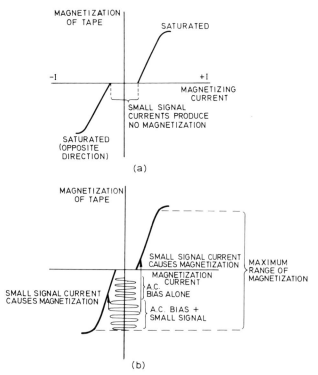

Fig. 12.4. Use of a.c. bias; (a) With no bias. (b) Effect of bias

from the head before any significant reversal of the field occurs. The higher the tape speed, the better the quality of the recording. The earlier tape recorders operated at 19cm per second, with 38cm per second as the standard speed for high-quality recordings. Cassette recorders operate at a much slower speed (4.76cm/s) but use tape heads in which the gap between poles is very narrow. This combines good quality with economy in the use of tape, though it leads to the problem of wear at the tape-head and the need to keep the head scrupulously clean.

One of the features of magnetic materials is that they are not magnetised at all unless the magnetising field exceeds a certain minimum strength. This means that signals of low level are not recorded but the problem is overcome by the use of an *a.c. bias* signal (Fig. 12.4). This signal is mixed with the audio signal at recording. It brings the strength of the magnetic field up into the range at which recording can occur. Then even the smallest of audio signals will be registered. Since the a.c. bias signal is in the ultrasonic range, usually 90kHz, this signal is not heard on reproduction.

Erasing

One of the advantageous features of magnetic tape is that recordings can be erased and the tape used again for other recordings. It also allows corrections to be made to existing recordings. Erasing consists in disarranging the domains, so that they come to lie in random directions, as in an unrecorded tape. The method used is to subject the tape to a strong alternating field and gradually reduce the strength of the field to zero. Domains vary in the strength of field needed to change their alignment. When the field is strong, all domains are affected at each changes of direction of the field but, as the strength of the field is reduced, more and more domains become unaffected and take up random directions. Eventually all are randomly orientated. Most tape recorders have a special erasing head. This is fed by the same a.c. signal that is used for the bias, though at greater strength. The gap of the erasing head is large so that its field extends over a considerable length of tape. As a region of tape passes away from the head it experiences a gradually decreasing alternating field and is demagnetised.

Magneto-resistive heads

Readers who remember the physics taught to them at school will recall that a magnet induces a current in a coil only if the magnet is *moving* relative to the coil. The same law holds in a tape-recorder. When a tape is being played, the magnetization of the tape is not enough in itself to induce a current in the playback head. The tape must be moving. The magnetic field in the region between the jaws of the tape head must be changing.

A magneto-resistive head works on a different principle. It depends upon the fact that the electrical resistance of some ferromagnetic alloys is affected by a magnetic field. Only a thin strip of the alloy is needed to detect a field, so heads can be made very small. This type of head does not require the tape to be moving. If required it can read the signal at any point on a stationary tape. The design can also be adapted for reading the content of bubble memories (p.167).

Video tape recorders

The technology of tape recording soon improved to such an extent that it became possible to record video signals as well as audio signals. Video signals include frequencies up to 5MHz, which means that a very high tape speed is required. Given an ordinary audio tape-recorder head, the tape speed would need to be over 60 *metres* per second. Other methods of recording and reading the tape have been devised. In the *transverse scan* recording format, the head (or heads in some systems) is carried on a rotating drum. The tape is wide (Fig. 12.5a), and the more-or-less transverse segments of track are compactly aligned on the tape. Sophisticated electronic circuits are required to switch between the heads (if more than one is used) and to

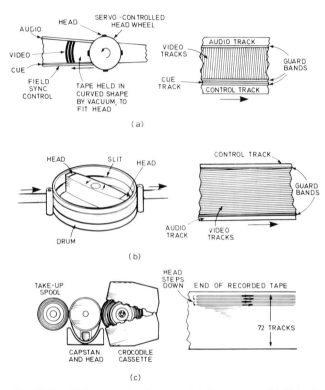

Fig. 12.5.   Video scanning formats.   (a) Transverse.   (b) Helical.
(c) BASF LVR

merge the signals from the segments of track into one continuous
signal on replay. Such systems are widely used in broadcasting.

  The scanning system developed for home video is the *helical
scan.* Tape is wrapped around the head as it passes from reel to
reel. The signal is recorded along a series of diagonally aligned
tracks (Fig. 12.5b).

Longitudinal scanning as in an ordinary audio tape recorder, was
the first method to be tried. It was soon abandoned because of
the difficulty of controlling tape at high speed and the fact that

impossibly long tapes were required to record a programme of even moderate length. Some of the lastest machines have reverted to longitudinal recording, though using improved formats and improved heads. The BASF format (Fig. 12.5c), uses an 8mm tape running at 4 metres per second. It records 72 parallel tracks, each 2½ minutes long. The tape runs backward and forward from reel to reel and the head steps down from one track to the next at each reversal of direction. The Toshiba format uses a continuous loop of 12.5 mm tape with 300 parallel tracks, each lasting 24 seconds.

There are many recording formats of each kind, each favoured by a different manufacturer. Each has its own special features so tapes recorded on one make of recorder cannot be played back on a machine of a different make.

Video discs

Video signals, together with their audio signal can be recorded in frequency-modulated form on a grooved disc similar to the ordinary audio disc. The disc must rotate at high speed (up to

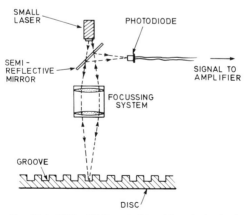

Fig. 12.6. Philips VLP optical head for playback of video discs

1800 rev/min) and a special stylus is required, but such systems
have proved workable. The major problem is that once the disc
is damaged it becomes unplayable. Manufacturers have therefore tried to produce systems in which there is no mechanical
contact between disc and the play-back head, so reducing the
risk of damage and eliminating noise. In the Philips system the
disc has a reflective surface. A beam of light from a small laser is
focused on the floor of the spiral track (Fig. 12.6). There the
video signal is represented as a row of pits of varying depth. The
reflected beam is detected by a photodiode (p.85). As the disc
rotates the head is moved across the disc, reading the signals
reflected from the floor of the track.

   In another contact-free system, the disc is electrically conductive and acts as one plate of a capacitor. The other plate is a
small one, located on the playing head. As the disc rotates the
distance between disc and head varies owing to tiny pits cut in
the disc. The capacitance between disc and head is varied at high
frequency, so providing the video signal.

Video cinema

A feature film on video tape costs less than a 35mm optical print,
and the cost of equipping a cinema with tape-player and
projector is less too. A number of cinemas are now equipped
with video-tape systems. The chief concern is how to provide a
picture that is large enough to be viewed by an audience, yet has
brightness and definition comparable to that produced by an
optical projector. The light guide projector employs three tubes
of the kind shown in Fig. 12.8. Each produces its own coloured
image (red, green or blue) and these images are superimposed
on the screen to form a single full-coloured image. Since the light
from three tubes is combined, a bright image is obtained.

   The optical cinema projector owes its brilliant image to the
exceedingly bright lamp that is its source of light. The phosphor
image on a video tube can not match this level of brilliance. One
solution to this problem is found in the Eidophor tube (Fig.
12.8). Its source of light is a high-powered xenon lamp. The light

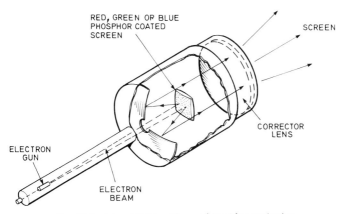

Fig. 12.7. Advent light guide, used in video projection

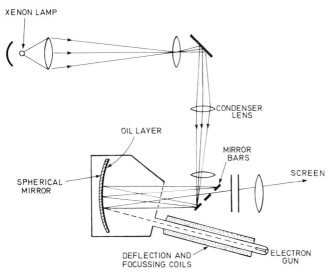

Fig. 12.8. Eidophor video projection tube

is focussed on to a mirror that is covered with a layer of oil. When the electron beam is not scanning, the surface of the film of oil is smooth; the light reflected from the mirror is blocked by the bars and does not reach the screen. When the film of oil is scanned, its surface is disturbed. The higher the intensity of the beam, the greater the disturbance and more light is reflected to pass between the bars to the screen. For colour projection, three Eidophor tubes are used, one for red, one for green and one for blue. The projected images are superimposed on a screen to obtain a full-colour picture.

Digital recording

Digital recording brings the precision of the binary system to both audio and video recording. As was explained on (p.148), the coding of information in binary form allows it to be manipulated at high speed, with absolute accuracy and with relatively simple and robust circuits. In converting an analogue signal (audio or video) into digital form we sample it at regular intervals (Fig. 12.9). Then its value at each interval is encoded in binary form. The sequence of binary numbers at the bottom of the figure is a coded representation of the wave-form. The greater the number of bits in the code, the greater the precision of the recording. The frequency of sampling depends upon the type of signal. For audio frequencies up to 20kHz we need to sample at twice this frequency, i.e. at 40kHz. Digital signals consist of '1's and '0's which can be represented on tape by just two magnetization levels (see Fig. 12.3, p.216). Compare this with analogue recording in which we require an infinite number of levels from zero almost to saturation in either direction. Recording and playing circuits can be made to respond perfectly to digital signals, so there is no loss of quality during recording or play-back. Further, since the production of recordings often involves the re-recording and mixing of previously recorded material, there are usually several stages of recording and replaying between the original performance and the final emergence of sound from the loudspeaker. At each of these stages

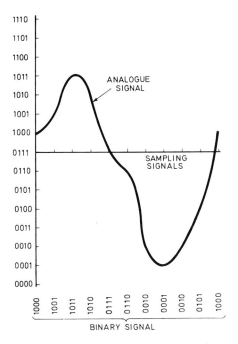

Fig. 12.9. Digital representation of an analogue
audio or video signal

there is distortion and loss of quality–even with the best of equipment. With digital recording the umpteenth re-recording has the same high quality as the first recording.

The advantages of digital techniques also apply to the transmission of signals over long distances. System X, developed by British Telecom in association with a number of British telecommunications firms, uses digital coding to improve the quality of telephone transmission. The system is controlled by computer so that various kinds of service may be performed automatically –for example, it can transfer calls to another telephone if you are away for the day. The system switches all signals by electronic logic circuits, so that the old-fashioned mechanical switches and relays are no longer required. Reliability is improved by about 20 times, yet a System X exchange is only one twentieth the size of a mechanical exchange that would handle the same number of calls. All of these developments depend entirely on the existence of electronic logic devices of the kind described in Chapter 8.

13 Electronics in Industry

The use of electronics in industry has greatly increased in recent years. Electronic techniques can be used to measure thickness, temperature and strain, for weighing and counting, to control drying processes, detect radiation, measure flow, control electric motors and so on. Computers can take the place of supervisory staff to control complex operations and may be used to work out the cost and possibly the profitability of proposed industrial schemes.

This chapter gives a few examples of the industrial use of electronics to show the useful role that electronics plays in modern industry.

Electronic counting

There are a number of ways in which electronic techniques can be used in factories where large quantities of small items, for example small parts, have to be counted.

One method uses an ordinary microphone as shown in Fig. 13.1. The small metal parts drop singly on to a platform under which a microphone is situated. The microphone records the sound as each item hits the platform prior to slipping away down the chute to a collector or to a moving belt to be carried away for packaging. The microphone output is connected to an amplifier the output of which is fed to an electromechanical counter which records the number of sounds picked up by the microphone.

A more sophisticated method which may be used where larger objects are to be counted, are fragile or cannot be handled uses a photocell as shown in Fig. 13.2. The photocell is placed opposite a light and as objects pass they interrupt the light beam. When the light in the photocell is thus interrupted the photocell is inactivated. Each time this occurs the photocell applies a signal to the amplifier which produces a pulse that is fed to an electromechanical or electronic, e.g. a digital, counter. It can be arranged that the counter provides an output signal after a certain number of objects have been counted, the output signal being used to initiate some further action such as stopping the conveyor belt.

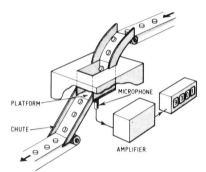

Fig. 13.1. Counting small metal articles by using a microphone and counter

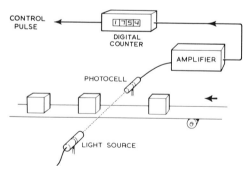

Fig. 13.2. Using a photocell to count packages

As an extension of this method, it is possible to use the counting rate to control the time of some process such as drying or curing. For example, if electronic potted circuits are counted after passing through a curing oven to harden the potting resin, the rate at which they are counted is a function of the time they spend in the curing oven. If this information is correlated with the oven temperature it is possible to ensure, by means of suitable control equipment, that each potted circuit spends the correct time in the oven.

Measuring flow

Measuring the flow of liquids in pipes with no restriction to the flow of the liquid can be achieved by means of the arrangement shown in Fig. 13.3. A coil is wound around the pipe carrying the fluid and a large magnet is placed across it. The liquid, which

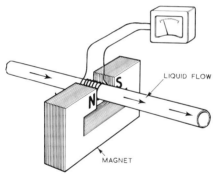

Fig. 13.3. Measuring the flow of liquid in a pipe

must of course be electrically conducting, flows through the magnetic field so that a current is induced in it. This is coupled to the pick-up coil which provides an indication on a meter. The deflection on this meter is directly proportional to the amount of liquid flowing and thus the meter reading is a measure of the rate of flow.

Another type of flowmeter, which can be used when the liquid is non-conducting, is shown in Fig. 13.4. This is a more expensive equipment and uses ultrasonics in measuring the flow. Two barium titanate crystals are placed in the pipe diagonally. Ultrasonic energy is fed to one crystal at a time while the other is used to receive the ultrasonic energy. In this way the time taken for a pulse of ultrasonic energy to travel between the crystals can

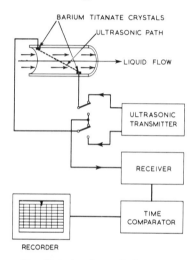

Fig. 13.4. An ultrasonic flowmeter

be measured. If the pulse is fired against the fluid flow it will take longer to reach the receiver crystal than if it is fired downstream. The difference in these times is directly proportional to the rate of flow. The pulses are fired alternately from each crystal, and picked up by the other, the times being compared in an electronic comparator circuit. The final output, which is proportional to the flow rate, may be recorded.

Moisture content

A relatively simple technique can be used to test materials like cloth and paper for moisture content. The material to be tested is passed over rollers between two metal plates which act as the

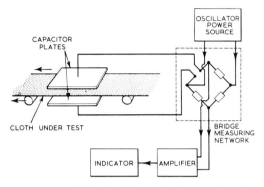

*Fig. 13.5. Method of electronically measuring the
moisture content of cloth*

plates of a capacitor. The material is thus the equivalent of the
dielectric of a capacitor.

If the cloth contains moisture the value of the capacitance will
be increased, and vice versa. If this capacitor is made one arm of
a measuring bridge then the amount of moisture present in the
cloth will be converted into an electrical signal which can be used
to give an indication or warning or take some action such as
increasing the temperature of driers. Fig. 13.5 shows a possible
arrangement for checking moisture content of cloth passed over
rollers.

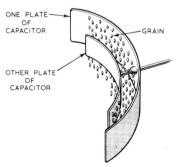

Fig. 13.6. Grain drier sensor

In agriculture a similar method is used to dry out moist grain. A large specially shaped capacitor is situated in the mass of grain which is passing through the drier. The change in capacitance forms a signal which is applied to a servomechanism. This alters the rate at which the grain travels through the drier to ensure that it is evenly dried. The capacitor used in this application is shown in Fig. 13.6.

Ultrasonics in industry

Sound waves are longitudinal disturbances in a medium such as air or water. Electromagnetic waves differ in being transverse in character. The energy changes of transverse waves occur in a plane vertical to the direction of movement, rather like ripples on a pool of water when a stone is thrown in, whereas sound waves make alternate rarefactions and compressions of the medium in which they travel, which of course must be elastic in order to transmit them.

Ultrasonic waves are the same as sound: they are not, as some people imagine, like radio waves. The confusion arises because they are high in frequency–they may even be in the megahertz region. Even so they are totally unlike the radio frequency waves. The term ultrasonic means 'above sound' and that is precisely what they are. Many people cannot hear any sound with a frequency higher than about 8000 Hz so that to them anything above this value would be ultrasonic. Generally speaking, however, it is usual to regard ultrasonic waves as being those above about 16 000 Hz.

To produce an ultrasonic disturbance in water or a similar medium some form of transducer is needed. This may be made of barium titanate–a material which has piezo-electric properties, that is, it oscillates mechanically when electrical oscillations at the right frequency are applied across it, and vice versa.

A typical ultrasonic transducer is shown in Fig. 13.7. The leads are coupled to a source of radio frequency energy of appropriate power. A transducer such as that shown needs about as much power to operate it as a large electric light bulb. Ultrasonic

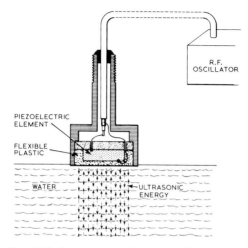

*Fig. 13.7. Transducer producing an ultrasonic
disturbance in water*

oscillations are transmitted into the medium, in this case water,
and if the transducer is correctly shaped the energy will be
transmitted as a beam.

Detecting flaws with ultrasonics

In the iron founding industry ultrasonics has assisted in making
quality testing a much more exact task. In the past, large iron
castings weighing many hundreds of pounds were sold to custom-
ers for machining, a process that might take several weeks, and it
was quite possible that at the end of this time, while a final cut
with a milling tool was being made, a large hole would be
revealed in the block due to faulty casting. This meant not only a
waste of the casting but serious loss of time. Today, however,
although flaws still occur in large castings–in fact it is impossible
to guarantee that they will not sometimes occur–they can be

detected before the casting is sold so that loss of time and money is avoided.

This quality checking is done by an ultrasonic flaw detector. The transducer head is coupled to the casting as shown in Fig. 13.8, using some water-bound cellulose paste, and the oscillator producing the energy to power the transducer is switched on.

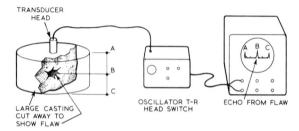

Fig. 13.8. Ultrasonics used to check for flaws in metal castings

This oscillator is pulsed (as in radar) so that it produces short sharp output pulses at the rate of perhaps several hundreds per second, each pulse lasting for perhaps a few milliseconds. The pulses are fed to the transducer and transmitted though the casting, and echoes will return along the path of the beam. These echoes are picked up by the transducer, reconverted to electric signals, and can be displayed on an oscilloscope. Echoes occur whenever the ultrasonic beam encounters an alternation, e.g. a flaw, in the medium through which it is being transmitted. In this way the presence of flaws will show up as echo signals on the oscilloscope.

The same transducer head acts as both the transmitter and receiver of the ultrasonic energy: after each transmitted pulse the head is electronically switched over to 'receive' to await the return pulses from the casting. Fig. 13.8. shows a typical set of echoes from a casting, echo A being caused by the interface between the transducer and the casting, B by a flaw in the casting and C by the interface between the casting and the surface on which it stands.

Automation in industry

Electronic techniques have made possible great advances in automation in the last few years and there is today hardly a plant in the country without some automated process. In the chemical industry for example some large refining plants are run for months with only a handful of staff, most of the processes being automated.

Fig. 13.9 shows in outline a basic automatic control system. Suppose that we wish a conveyor belt to travel at a certain speed. This is represented in Fig. 13.9 as the 'operation under control.'

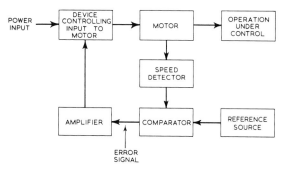

Fig. 13.9. Block diagram showing the basic elements of a closed-loop automatic control system

The conveyor belt is driven by an electric motor. To this is linked a device providing an electrical signal which is proportional to the motor's speed. This signal is compared by the comparator with a *reference signal* corresponding to the speed at which the motor should run to drive the conveyor belt at the desired speed. The comparator provides an *error signal* corresponding to the difference between the actual speed of the motor and its correct speed, and this error signal, after amplification, is used to control the power input to the motor so that the motor's speed is automatically compensated. This is called a 'closed-loop' feedback control arrangement: information on the actual operation under control is fed back and used to adjust it automatically.

There is, of course, always some slight error in the system because deviation cannot be corrected until after it has taken place, but this error can by careful design be kept very small.

Thyristors are electronic devices that are used frequently for the control of electric motors. Where a d.c. motor is powered by an a.c. supply after rectification, they can be used conveniently to provide the necessary rectification. In such cases the error signal can be used to control the gate of the thyristor used as a rectifier, thereby controlling the power supply to the motor.

Feedback control arrangements can be devised to control temperature, humidity, flow, pressure, weight–in fact almost any factor which can be electronically measured by means of a suitable transducer. Also, if the reference source is 'programmed,' i.e. takes the form of a series of different signals, then the operation under control can be made to perform a series of operations automatically. A programme can be provided by a computer or a pre-recorded tape, and used in the control, for example, of machine tools.

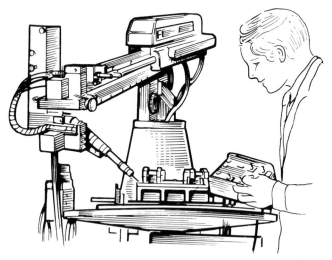

Fig. 13.10. A factory robot. The engineer is 'teaching' the machine to carry out a set routine, after which it will work automatically

Electronic systems of automation relieve man of the tedious tasks of measuring, assessing and correcting equipment, and can operate at much faster speeds.

Many factory jobs such as painting, welding and the assembling of machinery are performed by robots. Some of these are taken through the required routines by a human operator Fig. 13.10. After one such 'training' session the robot 'remembers' the movements required (for example, to spray the body of a car) and repeats the action on every subsequent occasion without need for human guidance. The essential features of robots is that they are *flexible*, they can be programmed to perform a wide variety of tasks so that the same robot can be used in different parts of the production line, or can be re-programmed when the product design is altered.

At this point in our account of automation technology we are going beyond the realm of electronics. The action of robots is controlled by software, and their seeming intelligence is the result of the skill of the programmer. Their reliability is the result of the skill of the engineer. These are aspects that it would be out of place to discuss here. But the electronic sensors used by the robots are those described in earlier chapters of this book. The electronic circuits of the robot are based on the transistors and logic gates we have already described. The beginner who has read this far in the book would find nothing new in the electronic aspects of these apparently human machines.

Index